Johannes Reuther

Frustrated Quantum Heisenberg Antiferromagnets

Johannes Reuther

Frustrated Quantum Heisenberg Antiferromagnets

Functional Renormalization-Group Approach in Auxiliary-Fermion Representation

Südwestdeutscher Verlag für Hochschulschriften

Impressum/Imprint (nur für Deutschland/only for Germany)
Bibliografische Information der Deutschen Nationalbibliothek: Die Deutsche Nationalbibliothek verzeichnet diese Publikation in der Deutschen Nationalbibliografie; detaillierte bibliografische Daten sind im Internet über http://dnb.d-nb.de abrufbar.

Alle in diesem Buch genannten Marken und Produktnamen unterliegen warenzeichen-, marken- oder patentrechtlichem Schutz bzw. sind Warenzeichen oder eingetragene Warenzeichen der jeweiligen Inhaber. Die Wiedergabe von Marken, Produktnamen, Gebrauchsnamen, Handelsnamen, Warenbezeichnungen u.s.w. in diesem Werk berechtigt auch ohne besondere Kennzeichnung nicht zu der Annahme, dass solche Namen im Sinne der Warenzeichen- und Markenschutzgesetzgebung als frei zu betrachten wären und daher von jedermann benutzt werden dürften.

Verlag: Südwestdeutscher Verlag für Hochschulschriften GmbH & Co. KG
Dudweiler Landstr. 99, 66123 Saarbrücken, Deutschland
Telefon +49 681 37 20 271-1, Telefax +49 681 37 20 271-0
Email: info@svh-verlag.de

Zugl.: Karlsruhe, Karlsruher Institut für Technologie, Diss., 2011

Herstellung in Deutschland:
Schaltungsdienst Lange o.H.G., Berlin
Books on Demand GmbH, Norderstedt
Reha GmbH, Saarbrücken
Amazon Distribution GmbH, Leipzig
ISBN: 978-3-8381-2731-6

Imprint (only for USA, GB)
Bibliographic information published by the Deutsche Nationalbibliothek: The Deutsche Nationalbibliothek lists this publication in the Deutsche Nationalbibliografie; detailed bibliographic data are available in the Internet at http://dnb.d-nb.de.

Any brand names and product names mentioned in this book are subject to trademark, brand or patent protection and are trademarks or registered trademarks of their respective holders. The use of brand names, product names, common names, trade names, product descriptions etc. even without a particular marking in this works is in no way to be construed to mean that such names may be regarded as unrestricted in respect of trademark and brand protection legislation and could thus be used by anyone.

Publisher: Südwestdeutscher Verlag für Hochschulschriften GmbH & Co. KG
Dudweiler Landstr. 99, 66123 Saarbrücken, Germany
Phone +49 681 37 20 271-1, Fax +49 681 37 20 271-0
Email: info@svh-verlag.de

Printed in the U.S.A.
Printed in the U.K. by (see last page)
ISBN: 978-3-8381-2731-6

Copyright © 2011 by the author and Südwestdeutscher Verlag für Hochschulschriften GmbH & Co. KG and licensors
All rights reserved. Saarbrücken 2011

Contents

1 **Introduction** — 1

2 **The J_1-J_2 Heisenberg Model** — 5

3 **Auxiliary Fermions** — 9

4 **Mean-Field Theory** — 13
    4.1 Hartree Approximation — 13
    4.2 Random-Phase Approximation — 16

5 **Finite Pseudo-Fermion Lifetime** — 21
    5.1 Hartree Approximation — 22
    5.2 Random-Phase Approximation — 24
    5.3 The Spectral Width in Diagrammatic Approximations — 28

6 **The Functional Renormalization Group: Implementation for Spin Systems** — 31
    6.1 General FRG Formalism — 32
    6.2 FRG and its Implementation for Heisenberg Systems — 36
    6.3 Static FRG Approach — 47
    6.4 Conventional Truncation Scheme — 50
    6.5 Katanin Truncation Scheme — 52
    6.6 Dimer and Plaquette Order — 60

7 **Application to Further Models** — 63
    7.1 The J_1-J_2-J_3 Heisenberg Model — 63
    7.2 The Heisenberg Model on a Checkerboard Lattice — 67
    7.3 The Anisotropic Triangular Heisenberg Model — 71
    7.4 The Heisenberg Model on a Kagome Lattice — 77
    7.5 The Heisenberg Model on a Honeycomb Lattice — 78
    7.6 The Kitaev-Heisenberg Model — 84

8 **Pseudo-Fermion FRG Including Magnetic Fields** — 91
    8.1 Modifications of the Formalism — 91
    8.2 Hartree- and Random Phase Approximation — 94
    8.3 Full One-Loop FRG Scheme — 96

9 **FRG at Finite Temperatures** — 99

9.1 Modifications of the Formalism 99
9.2 Results for the J_1-J_2 model 101

10 Limitations of the FRG: Lower Dimensions   105

11 Conclusion and Outlook   109

A The Popov-Fedotov Technique   113

B Flow Equations for the Two-Particle Vertex   115

C Symmetries of the Two-Particle Vertex in the Transfer Frequencies   119

1 Introduction

The quantum theory of magnetism with its richness of related phenomena has always been a fascinating subject in condensed matter physics. Starting from the microscopic picture of localized magnetic moments arranged in some kind of lattice, a lot of interesting physics emanates from correlation effects mediated by the interaction. In this context, a fundamental and extensively studied system is the Heisenberg model, describing isotropic two-body exchange interactions between moments on nearest neighbor sites (or, in its generalized version also between sites being further apart). Despite the simplifying assumption that only the spin degree of freedom is relevant while the charge is frozen, the physics contained in the family of Heisenberg models is of enormous variety and opens the door to a wide range of applications as described below. The most interesting situation is encountered in the extreme quantum limit where the magnetic moments carry spin-1/2 and, moreover, in the case of antiferromagnetic interactions since such systems are strongly affected by quantum fluctuations at low temperatures, giving rise to exotic quantum states. Together with the effect of frustration in the form of competing spin interactions this is the general setup for the investigations in this thesis.

While Heisenberg models have often been in the focus of condensed matter theory, partially in very different contexts, the motivation for studying these systems has changed considerably during the past decades. Proposed in 1928 by Heisenberg [55] and Dirac the exchange interaction represented a new mechanism to describe the correlations in ferromagnetic materials, which was not possible on the basis of magnetic dipole-dipole interactions as they are several magnitudes too small to explain the observed Curie temperatures. Although the Heisenberg model is not directly applicable to itinerant ferromagnets like Fe, Co, Ni, the underlying idea of exchange interactions proved to be correct. In fact, in the early 60's some magnetic, isolating rare earth and transition-metal compounds such as EuO [77] and $RbMnF_3$ [133] turned out to be perfect realizations of the nearest-neighbor Heisenberg model.

After the discovery of high-T_c superconductivity in 1986 [17, 37] the Heisenberg model gained renewed interest. The two-dimensional CuO-planes, which represent a typical feature of all cuprate superconductors, are in fact well described by a nearest-neighbor spin-1/2 Heisenberg model explaining the antiferromagnetic state of the undoped parent compound. When a small concentration of holes is doped into the CuO-planes, magnetic order is rapidly destroyed, giving way to a non-magnetic pseudo-gap state and, upon further doping, to superconductivity [68, 134]. Early theories on high-T_c superconductivity have been strongly influenced by the physics of pure spin models: Anderson proposed that a non-magnetic resonating valence-bond (RVB) state, which has first been introduced in the context of two-dimensional antiferromagnetic Heisenberg models [8], forms the fundamental basis on which the theory of high-T_c superconductivity should be built [9]. It is argued that there is a direct correspondence between the singlet pairs of the insulating state and the charged superconducting pairs when the insulator is sufficiently doped. Although this idea has been considered by many authors since then, there is no conclusive answer to the question of the role of a spin liquid state for high-T_c superconductivity. Today, there is at least general agreement that the physics behind the phase diagram of the cuprates is the physics of the doping of a Mott insulator which is believed to be captured by the t-J model. The latter in turn reduces to the Heisenberg model at half filling.

These early studies have raised the question, however, under which conditions quantum fluctuations are strong enough to destroy long-range order in Heisenberg systems. Thermal fluctuations are important as well, especially since they suppress long-range order in two dimensions at any finite temperature, but their role is relatively well understood. By contrast, quantum fluctuations operate in a much more complex way: They may suppress long-range order, but may at the same time lead to novel ground states known under the labels "spin liquid" (as the aforementioned RVB state) and "valence-bond solid" (VBS). The time after the discovery of high-T_c superconductivity was characterized by a huge number of studies on many different two-dimensional Heisenberg-like systems, some of which are also investigated in this thesis. Frustration effects, either by competing spin interactions or due to special geometric arrangements have always been of particular interest, especially as it turned out that upon tuning the interactions or the lattice anisotropy many systems may be driven into a phase without magnetic long-range order. Very often, such discoveries came along with new methodological developments of both, analytical and numerical type. Nevertheless, the adequate treatment of spin systems in the thermodynamic limit remains a complicated task such that until now each approach suffers from some kind of drawback. Especially the identification of the nature of non-magnetic phases turned out to be very challenging: While for some systems a valence-bond solid ground state, i.e., a state with hidden long-range order in the form of some type of dimerization, is clearly favored [48, 74, 84], a disordered spin liquid has not yet been detected in a completely unbiased way.

In the 90's, accompanied by the progress in the understanding of Heisenberg models, also the theory of quantum phase-transitions has experienced renewed interest [104]. Even more recently, the notion of "deconfined quantum criticality" [110, 111] gained much attention as a mechanism to explain how two differently ordered phases may be connected by a continuous phase transition, which would contradict the common Ginzburg-Landau-Wilson paradigm.

Another fascinating perspective in the context of quantum spin systems concerns topological quantum computation, which has recently become a new field in condensed matter theory. By means of two-dimensional excitations called anyons (i.e., particles which are neither fermions nor bosons) as topologically non-trivial quasiparticles whose worldlines form a braid, a realization of quantum memory has been proposed which is protected from decoherence [70]. Although the spin systems that are known to possess anyonic excitations involve anisotropic spin couplings or even four-body interactions and are therefore not of Heisenberg type, a related system will also be studied in this thesis, see Section 7.6.

From our viewpoint there are several reasons to study Heisenberg models: As many aspects of our approach presented in this thesis are associated with new developments, a first motivation is of purely methodological type. Secondly, as the next step, we like to contribute to the search for novel non-magnetic ground states in highly frustrated spin models, which has been a long-standing problem for so many years. Finally, in order to make contact to actual experiments we aim to investigate models for materials which are of current interest (see Sections 7.3, 7.4 and 7.6).

In this thesis we develop new analytical and numerical methods for calculating ground-state properties of a large class of spin models on the basis of infinite resummations of perturbation theory in the couplings. To this end we use a representation of the spin operators in terms of pseudo fermions [1]. One motivation for using a fermionic representation rather than a bosonic representation is the available experience in describing spin liquids or dimerized spin-singlet states with fermions, mainly within large-N and mean-field approaches (see e.g. Refs. [3, 10, 22, 101]). On the other hand, pseudo-fermion representations have hardly been used to study magnetic ordering phenomena [61]. Although a large body of results of numerical studies of these models is available, analytical approaches starting from a microscopic Hamiltonian are rare. We use a newly developed implementation of the functional renormalization group (FRG) method [67, 107] applied to interacting quantum spin models. Auxiliary particle representations of spin operators are sometimes viewed with suspicion, as they are conceived to be fraught with uncontrolled approximations regarding the projection unto the physical sector of the Hilbert space necessary in those spin representations. Here we are using an exact method of projection onto the physical part of Hilbert space that works even on the lattice.

Applying our method to frustrated spin systems, we show that the FRG based on pseudo fermions is capable of giving results in very good agreement with results obtained mainly by purely numerical means. Furthermore, we demonstrate that the approach is able to (i) treat large system sizes of $\mathcal{O}(200)$ sites, (ii) is applicable to arbitrary frustrated lattice geometries and two-body bare interactions, (iii) naturally allows to compute the magnetic susceptibility as the canonical outcome of the RG, and (iv) hence provides an unbiased calculation from first principles that allows comparison to experiment.

This thesis is organized in the following way: **Chapter 2** introduces the J_1-J_2 Heisenberg model which provides a suitable testing ground for various approximation schemes applied in the subsequent chapters. The auxiliary-fermion representation and the projection schemes onto the physical Hilbert space are presented in **Chapter 3**. Simple mean-field approximations are discussed in **Chapter 4** where we demonstrate that these approaches are not able to capture frustration effects but rather reproduce classical results. To this end in **Chapter 5**, we introduce a phenomenological pseudo-particle lifetime that mimics quantum fluctuations. The results on the magnetization, susceptibility, dynamical spin-structure factor and spatial spin correlations show that in a certain parameter range for this lifetime, the correct phase diagram is obtained.

After these preliminary considerations the main methodological part of the thesis, given by **Chapter 6**, is devoted to FRG. This method enables us to calculate the auxiliary particle damping rather than treating it as an input of the approximation. To start with, we give a brief review of the FRG approach in general, especially its derivation in the Feynman path integral formalism. Thereafter, in Section 6.2, the FRG implementation specific to Heisenberg spin systems is presented. All new developments that are required to describe spin systems within FRG are contained in this section. After a brief discussion of static FRG schemes in Section 6.3, the non-trivial issue of how the hierarchy of FRG equations should be truncated is discussed in the next two sections: In Section 6.4 it turns out that within a pure one-loop formulation, quantum fluctuations are not sufficiently accounted for, such that on application to the J_1-J_2 model the expected non-magnetic intermediate phase is not found. We trace this deficiency of the one-loop approximation to the neglect of higher order contributions, with the consequence that not even the dressed RPA scheme is reproduced. As shown by Katanin [67] the latter problem may be remedied by using a modified single-scale propagator, thus including certain three-particle correlations with non-overlapping loops. Section 6.5 demonstrates that upon using the Katanin truncation scheme we find a phase diagram in good agreement with results from numerical methods. The chapter closes with the discussion of a scheme that allows to estimate dimer fluctuations in paramagnetic phases, see Section 6.6.

Subsequent to Chapter 6, which has been mainly devoted to technical issues, **Chapter 7** presents the FRG results for further spin systems. We demonstrate that the FRG with pseudo fermions in conjunction with the Katanin truncation is not only capable to describe the J_1-J_2 model but also gives correct results for more complicated systems like the J_1-J_2-J_3 square lattice model (Section 7.1), the Heisenberg model on a checkerboard- (Section 7.2), anisotropic triangular- (Section 7.3), Kagome- (Section 7.4) and honeycomb-lattice (Section 7.5) and finally the Kitaev-Heisenberg model (Section 7.6).

The next three chapters briefly present certain extensions of the FRG approach as it has been applied so far: In **Chapter 8** we modify the FRG such that SU(2) broken flows under the influence of external magnetic fields may be studied. On a pure meanfield level our results are in agreement with the general notion of symmetry breaking or linear response, see Section 8.2. While symmetry breaking by small magnetic fields is also well described within the full FRG scheme, the unbiased detection of non-magnetic phases turns out to be rather difficult on the basis of that approach. The FRG at finite temperatures is discussed in **Chapter 9** where we show that well controlled calculations can be performed at least at high enough temperatures. Since that approach allows us to measure the fulfillment of the pseudo-fermion constraint directly, we obtain the important result that the average projection used within our zero-temperature FRG scheme has been justified, see Section 9.2. **Chapter 10** contains a discussion on zero- and one-dimensional spin systems. There we illustrate that due to an overestimation of magnetic order in lower dimensions our method is most suitable for 2D spin systems. Finally, in the concluding **Chapter 11** our results are collected.

2 The J_1-J_2 Heisenberg Model

A prototype for a frustrated spin system is the antiferromagnetic spin-1/2 J_1-J_2-Heisenberg model on a two-dimensional square lattice. This model, in the following briefly called J_1-J_2 model, has been extensively studied for more than 20 years in order to understand the effect of competing spin interactions. In addition to the nearest neighbor interaction J_1, this model features a next nearest-neighbor interaction J_2, such that the Hamiltonian of the system is given by

$$H = J_1 \sum_{\langle i,j \rangle} \mathbf{S}_i \cdot \mathbf{S}_j + J_2 \sum_{\langle\langle i,j \rangle\rangle} \mathbf{S}_i \cdot \mathbf{S}_j . \tag{2.1}$$

Here the first term denotes a sum over all nearest-neighbor sites i, j and the second term is a sum over all next nearest-neighbor pairs i, j of the underlying two-dimensional square lattice, see Fig. 2.1. Since we are interested in the antiferromagnetic model only, both coupling constants are assumed to be positive, $J_1, J_2 > 0$. The only tuning parameter of the model is the ratio of these interactions which we define as $g = \frac{J_2}{J_1}$. There are still plenty of open questions concerning the ground-state phases and the corresponding phase transitions.

We begin the discussion of the ground-state properties in the classical large spin limit ($S \to \infty$), i.e., we consider the variables $\mathbf{S}_i$ as (classical) vectors with a fixed length. The phase diagram is then easily determined: For $J_2 = 0$, $J_1 > 0$ the spins are Néel-ordered (↑↓↑↓) which is energetically favored up to $g = 0.5$. However, in the opposite limit $J_1 = 0$, $J_2 > 0$ ($g \to \infty$) the two sublattices of the square lattice decouple such that on each sublattice the spins arrange in a Néel pattern. This is the so called collinear order configuration which ranges down to $g = 0.5$ where it meets the Néel phase in a first-order phase transition. As long as we consider the large spin limit, there is still the freedom to choose the angle between the interpenetrating Néel configurations of the collinear phase, which results in a large degeneracy of the classical ground state. In other words, one can rotate the two sublattices against each other. As a consequence of the classical character, the local magnetic moments in both phases are always given by their saturation values.

This situation changes when we turn to the spin-1/2 quantum case. The system is now influenced by quantum fluctuations which govern the relevant physics. It is a long known fact that for the bare nearest neighbor Heisenberg model ($g = 0$) quantum fluctuations reduce the Néel magnetization to about 60% of the saturation value [94]. Due to the equivalence of the cases $g = 0$ and $g \to \infty$ the same value for the magnetization also holds in the latter limit. Early on it has been found by spin-wave calculations, that the frustration further reduces the magnetic long-range order, especially near the classical

2 The J_1-J_2 Heisenberg Model

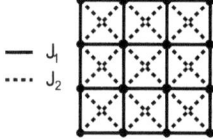

Figure 2.1: The J_1-J_2-Heisenberg model. Dots indicate the spins of the model.

transition point $g = 0.5$, even leading to a small parameter window without any magnetic order [36, 45]. Ever since, it has been the subject of a huge number of works using a variety of different methods to confirm this observation. However, even more than 20 years after the first studies of the J_1-J_2 model a clear statement about the exact positions of the transition points as well as a rigorous proof of the existence of this paramagnetic intermediate phase is still missing. Nevertheless, there is agreement that this phase approximately exists in a region $0.4 \lesssim J_1/J_2 \lesssim 0.65$ between the two ordered states [43, 46, 62, 109, 123, 131]. Another quantum effect concerns the collinear phase whose degeneracy is lifted compared to the classical large-spin limit: Quantum fluctuations fix the possible angles between the Néel configurations on the two sublattices such that they are either parallel or antiparallel [35]. Accordingly, the ground state in this phase is twofold degenerate: The alignment of parallel spins along rows and antiparallel spins along columns (corresponding to a magnetic wave vector $\mathbf{Q} = (0, \pi)$) or vice versa (corresponding to $\mathbf{Q} = (\pi, 0)$). This reduction of classical degeneracy due to quantum fluctuation is known as "order from disorder" [57].

Of more physical relevance than the exact positions of the transition points is the question of the nature of the non-magnetic intermediate phase which turned out to be a puzzling issue. Already in the early works there was disagreement about this phase either being a homogeneous spin liquid [36] or some kind of valence-bond solid (VBS) [41, 93]. In the latter case the spins form pairwise singlets which are spontaneously dimerized and therefore break e.g. lattice translation or lattice rotational symmetry. Also later when field-theory methods [93, 132], exact diagonalization [28, 30, 74, 46, 109], coupled cluster method [19, 43], series expansion [50, 123, 125, 131] and quantum Monte Carlo methods [27, 30] entered the field there was still no concensus about this question. One group claimed a VBS with a columnar dimerization [50, 123, 125, 131], the other group found a VBS with a dimerization that takes place on units of 2×2 plaquettes [30, 62, 131, 146], while a spin liquid could still not be ruled out [27, 126]. However the most recent mainly numerical work [43, 62, 125] clearly favors a VBS over a spin liquid. Evidence for a VBS has also been found in studies [116, 127] of a model of coupled spin chains [87], when the results are extrapolated to the isotropic J_1-J_2 model in the plane.

The question of the order of the quantum phase-transition has also attracted much attention. By now there seems to be agreement that the transition from the paramagnetic phase to the collinear configuration is of first order [43, 62, 123, 125, 131]. On the other hand the properties of the transition from the Néel phase to the paramagnetic phase are still highly controversial. Recent studies point to either a first order [125] or a second order transition [43, 62]. The latter scenario is the reason for a renewed interest in the J_1-J_2 model as it generally raises the question how two differently ordered phases may be connected by a continuous second order phase transition. If one assumes the non-magnetic phase to be a VBS which breaks rotational or translation symmetries but preserves the SU(2) symmetry, then a second-order transition to the Néel state which breaks the SU(2) and the translation symmetry but preserves the fourfold rotation symmetry is in contradiction to the usual Ginzburg-Landau-Wilson paradigm of phase transitions. In this case one would either expect a first-order critical point or a second-order transition that has to be described in terms of spinons as fractional excitations. As these spinons become deconfined at the critical point, this scenario is generally known as "deconfined quantum criticality" [110, 111].

The importance of the J_1-J_2 model is not only based on the theoretical issues stated above. It has also some relation to the cuprate superconductors where a small concentration of holes doped into the CuO-planes suffices to destroy the antiferromagnetic long-range order of the undoped system. The J_1-J_2 model as a simplified system for doped CuO-planes [60] describes the destruction of Néel order and therefore sheds light on the general question under which conditions long-range magnetic order may be destroyed.

Recently, this model has also found use for the vanadate compounds Li_2VOSiO_4 and Li_2VOGeO_4 which have been suggested as realizations of the J_1-J_2 model [32, 78]. For the former of these materials it is known that the next nearest-neighbor interaction is rather strong [32], i.e., $g \approx 1$. Indeed, experiments [78] reveal a collinear ordered ground state.

Even more recently the J_1-J_2 model has been invoked to account for the reduced magnitude of the ordered moment in the iron pnictides [42, 114, 115]. The undoped parent compound LaOFeAs described in terms of the J_1-J_2 model features a parameter ratio $g > 0.5$ [114] and correspondingly it has been shown in elastic neutron-scattering experiments that the system is collinear ordered. The frustration effects in conjunction with electron or hole doping further suppress the magnetic order and lead to the superconducting phase. The universally observed linear temperature dependence of the static, uniform magnetic susceptibility of these compounds has also been addressed in the framework of the J_1-J_2 model [145].

In summary, the J_1-J_2 model captures the main effects of competing spin interactions and at the same time features a relatively rich phase diagram comprising two magnetically ordered phases (Néel and collinear order) separated by a non-magnetic phase. Furthermore, the latter poses the challenging question of its precise nature. Therefore we consider the J_1-J_2 model as a good system to test different approximation schemes against each other [98]. In Chapter 7, however, we also turn to other frustrated spin systems.

3 Auxiliary Fermions

Auxiliary-particle techniques are widely used in the field of spin systems [13]. All approaches have in common that the spin operators are rewritten in terms of bosons or fermions which are easier to handle due to their canonical commutation relations. However, such spin representations are often accompanied with drawbacks of different kinds as they either require a complicated auxiliary-particle structure or are valid only in conjunction with particle-number constraints. In this chapter we present a brief overview of different auxiliary-particle techniques and introduce the Abrikosov pseudo-fermions in detail, since they form the basis of all approximations for frustrated spin systems in the following chapters. Furthermore, it is explained how the difficulties mentioned above are resolved in our approach.

Certainly, the most common auxiliary-particle approach is spin-wave theory [7, 75]. Starting from the bosonic Holstein-Primakov representation of spin operators, the essential step of approximation is the linearization of the representation which corresponds to the leading order in a $1/S$ expansion. Despite the fact that this scheme implements the large spin limit, spin-1/2 systems are often described surprisingly well on a qualitative level. For magnetically ordered phases this scheme has the advantage that in the linearized version the bosonic operators describe deviations from a spin-polarized state and are therefore directly associated to the magnons as the elementary excitations. As already pointed out, spin-wave theory is able to detect the paramagnetic phase in the J_1-J_2 model [36, 45] even though the method has a clear bias towards magnetically ordered phases.

In order to treat spin systems from the opposite viewpoint of non-magnetic valence-bond solids the bond-operator method has been developed [105]. Given a specific dimer covering of the lattice, the spin operators are expressed in terms of bosons which create (or annihilate) singlet and triplet states on these bonds. The corresponding Hamiltonian can then be diagonalized e.g. on mean-field level. However, the fact that only one singlet or triplet state exists on a bond imposes a local particle-number constraint on the bosons. This approach is particularly valuable for systems where the lattice structure itself suggests a certain dimer covering, e.g. the bilayer Heisenberg model [143].

Regarding the two-dimensional Hilbert space of a single spin one is tempted to use fermions for a spin representation. This can be done with a Jordan-Wigner transformation [138] where a spin-raising operator S^+ (spin-lowering operator S^-) is identified with a fermionic creation (annihilation) operator. The problem that these fermions do not satisfy the correct (bosonic) commutation relation of spin operators on different lattice sites is resolved by the introduction of an additional phase factor in the spin representation. This phase in turn requires the existence of an oriented path through the lattice and therefore the method cannot be straightforwardly extended to systems with dimensions higher than one. In this way the Jordan-Wigner transformation demonstrates the mapping of spin-chain models to spinless fermions in one dimension and hence constitutes an example that within the concept of auxiliary particles also exact calculations may be performed.

In this thesis we rewrite the spin operators in terms of Abrikosov auxiliary-fermions [1, 23, 24]. This representation requires two fermionic operators $f_\uparrow$ and $f_\downarrow$ for each lattice site,

$$S_i^\mu = \frac{1}{2} \sum_{\alpha\beta} f_{i\alpha}^\dagger \sigma_{\alpha\beta}^\mu f_{i\beta} \,. \tag{3.1}$$

Here σ^μ ($\mu = x, y, z$) are Pauli matrices, $\alpha, \beta = \uparrow, \downarrow$ are spin indices and i is the site index. Throughout the thesis we use units with $\hbar = k_B \equiv 1$. An important advantage of this representation compared e.g. to the Holstein-Primakov description is the simple quadratic form of (3.1) which allows for the application of Feynman-diagram techniques [86, 100]. By construction, the representation (3.1) satisfies the correct commutation relation of spin operators,

$$[S^\mu, S^\nu] = i\epsilon_{\mu\nu\eta} S^\eta \,. \tag{3.2}$$

However, the introduction of auxiliary fermions comes along with an enlargement of the Hilbert space: The basis set for a single site is now spanned by four instead of two states. In second quantization these states are given by,

$$|0,0\rangle \quad , \quad |0,1\rangle \quad , \quad |1,0\rangle \quad , \quad |1,1\rangle \,. \tag{3.3}$$

The numbers denote the occupation numbers of the "up" and "down" fermions, respectively. Applying the operator (3.1) to these states, one can easily show that the singly occupied states correspond to the usual spin-1/2 $|\uparrow\rangle$ and $|\downarrow\rangle$ states while empty or doubly occupied sites are unphysical (they have spin zero, $S = 0$). Therefore, the representation (3.1) is valid only in conjunction with the auxiliary-particle constraint

$$Q_i = \sum_\alpha f_{i\alpha}^\dagger f_{i\alpha} = 1 \,. \tag{3.4}$$

A convenient approximate approach is to replace the constraint $Q_i = 1$ by its thermodynamic average, $\langle Q_i \rangle = 1$. For a translation invariant state the latter conditions are identical at each site, such that only a single condition remains. Since the constraint amounts to removing two of the four states per site, it is on average equivalent to half-filling of the system, which in case of particle-hole symmetry is effected by applying a chemical potential $\mu = 0$ to the pseudo-fermion system.

An exact treatment of Eq. (3.4) is difficult because the constraint requires the fermion number to be fixed to 1 at each lattice site i individually. A simple projection scheme can be performed exactly in models like the Anderson impurity and Kondo model [2, 40] where only a single site carries a spin of the form (3.1). This scheme introduces a chemical potential λ_i for the site i. Taking the derivative of a physical observable with respect to the fugacity and performing the limitation $\lambda_i \to \infty$, the contribution in the

3 Auxiliary Fermions

physical subspace is projected out. However, since this method requires the limitations λ_i to be performed for each site independently, it is not applicable for lattice models.

By now the only method to handle the constraint exactly even for lattice systems has been proposed by Popov and Fedotov [91]. It amounts to applying a homogeneous, *imaginary-valued* chemical potential $\mu^{\text{ppv}} = -\frac{i\pi T}{2}$, where T is the temperature. Thus, within this scheme, the Hamiltonian H is replaced by

$$H \longrightarrow H^{\text{ppv}} = H - \mu^{\text{ppv}} \sum_i Q_i \quad . \tag{3.5}$$

Note that H denotes the Hamiltonian (2.1) using the representation of spin operators (3.1). Given a physical operator $\mathcal{O}$ (i.e., an arbitrary sum or product of spin operators) it can be shown (see appendix A) that the expectation value $\langle \mathcal{O} \rangle^{\text{ppv}}$, calculated with H^{ppv} and the *entire* Hilbert space, is identical to the physical expectation value $\langle \mathcal{O} \rangle$, where the average is performed with the original Hamiltonian H. The projection works by virtue of a mutual cancellation of the unphysical contributions of the sectors $Q_i = 0$ and $Q_i = 2$, at each site. It should be emphasized that although the Hamiltonian H^{ppv} is no longer hermitian, the quantity $\langle \mathcal{O} \rangle^{\text{ppv}}$ comes out real-valued. If on the other hand $\mathcal{O}$ is unphysical in the sense that it is non-zero in the unphysical sector, e.g., the operator $\mathcal{O} = Q_i$, the expectation value $\langle Q_i \rangle^{\text{ppv}}$ is meaningless and one has $\langle Q_i \rangle \neq \langle Q_i \rangle^{\text{ppv}}$.

This approach is applicable to spin models [44, 69] but cannot be extended to cases away from half filling. Although μ^{ppv} vanishes in the limit $T \to 0$, in principle, the exact projection with $\mu = \mu^{\text{ppv}}$ and the average projection with $\mu = 0$ are not equivalent at $T = 0$. This is due to the fact that the computation of an average $\langle \ldots \rangle^{\text{ppv}}$ does not necessarily commute with the limit $T \to 0$. Nevertheless it can be expected that in the model considered here both projection schemes are identical at $T = 0$. This can be understood with the following argument: Starting from the physical ("true") ground state, a fluctuation of the pseudo-fermion number results in two sites with unphysical occupation numbers, one with no and one with two fermions. Since these sites carry spin zero the sector of the Hamiltonian with that occupation is identical to the physical Hamiltonian where the two sites are effectively *missing*. Thus a fluctuation from the ground state into this sector costs the binding energy of the two sites which is of the order of the exchange coupling, even in the case of strong frustration. Negative binding energies are a generic property of spin systems since the ground state energy per site found in numerical studies is always negative (for the J_1-J_2 model see e.g. Refs. [43, 62, 123, 125, 146]). Consequently, at $T = 0$ pseudo-fermion number fluctuations are not allowed and it is sufficient to use the simpler average projection with $\mu = 0$. In most calculations we restrict ourself to this method. However, we again emphasize that at $T > 0$ both schemes will certainly differ.

Since the discovery of high-T_c superconductivity in 1986 [17, 37] the pseudo-fermion representation (3.1) has been extensively used mainly to study spin liquids or dimerized spin-singlet states in the t-J model within large N and mean-field approaches [3, 10, 16, 22, 76, 101, 135]. A special focus has been on the interplay between different ordered phases corresponding to different types of order parameters and mean-field decouplings of the Hamiltonian. Such mean-field amplitudes are given by the expectation values

$\langle f^\dagger_{i\alpha} f_{j\alpha}\rangle$ and $\langle f_{i\uparrow} f_{j\downarrow}\rangle$ where the first gives rise to the so called resonating valence bond [9] (RVB) or flux phase [3, 76, 135] while the latter leads to d-wave paired states [102, 135] and to superconductivity. These works aimed to understand to role of a spin-liquid in high-T_c superconductors at small doping. In contrast the pseudo-fermion representation has hardly been used to study magnetic ordering phenomena [61]. Unlike linear spin-wave approaches the pseudo fermions appear quadratic in (3.1) and therefore they are not directly associated with magnons as the elementary excitations in magnetic phases. Thus, it is a priori not clear how magnetic order may be detected with pseudo fermions in general. However, in recent years some experience concerning this issue has been obtained [23, 24]. In this thesis we aim to describe both, the ordering tendencies which are generally present in Heisenberg models as well as disorder effects which we generate through frustrating interactions.

The spin representation (3.1) also holds if the fermions are replaced by bosons (so called Schwinger bosons [12, 14]). One motivation for using the fermionic description rather than the bosonic representation is the aforementioned experience in describing spin liquids. Additionally, for a bosonic scheme the physical subspace forms a smaller sector of the total Hilbert space making the fulfillment of the constraint more complicated.

In the following we will formulate approximations in terms of resummed perturbation theory in the exchange couplings J_1, J_2 . The basic building blocks are the four-fermion interactions and the bare fermion Green's function in real space. Inserting the spin representation (3.1) into a Heisenberg Hamiltonian such as Eq. (2.1), a Hamiltonian quartic in the fermions is obtained. Due to the absence of quadratic terms, the bare Green's function in real space is given by

$$G^0_{ij,\alpha\beta}(i\omega) = \frac{1}{i\omega + \mu}\delta_{ij}\delta_{\alpha\beta} \quad , \quad \mu = -\frac{i\pi T}{2} \text{ or } \mu = 0\,. \tag{3.6}$$

without any self-energy contributions in the denominator. $\omega = (2n+1)\pi T$ are the fermionic Matsubara-frequencies. Note that in diagrammatic expansions the Green's functions remain strictly local, i.e., $G_{ij,\alpha\beta} = \delta_{ij} G_{i,\alpha\beta}$. The momentum dependence in correlators like the susceptibility is generated by the non-local exchange couplings.

In the next chapter we start with simple calculations on a mean-field level in order to demonstrate how in general magnetic long-range order is described within our pseudo-fermion representation.

4 Mean-Field Theory

4.1 Hartree Approximation

The most elementary approximation for a spin model is mean-field theory on the basis of spin operators. In our fermionic description it corresponds to the Hartree approximation shown in Fig. 4.1. The closed loop of the renormalized propagator represents the local magnetic moment which is proportional to the mean field, i.e., the Hartree self-energy. This self energy renormalizes the propagator which in turn feeds back to the local magnetic moment and thereby closes the self consistency.

Note that the Fock term is exactly zero, since the non-local exchange coupling connects two points of the same fermion line. The auxiliary-particle constraint forbids such terms because it assures that the fermion lines are local. Dropping the requirement of exact projection, one may allow for fermion hopping and make a mean-field ansatz with a non-vanishing Fock self-energy and non-local propagators. The corresponding symmetry-broken phase is the so called resonating valence bond (RVB) [9] or the flux phase [3, 76, 135] and the mean-field amplitude is given by the expectation value $\langle f_{i\alpha}^\dagger f_{j\alpha} \rangle$ already mentioned in Chapter 3. In principle one might also set up an approximation scheme using the pairing amplitude $\langle f_{i\uparrow} f_{j\downarrow} \rangle$ as mean-field parameter. This leads to superconducting d-wave paired states of the fermions [102, 135] and requires anomalous Green's functions in a diagrammatic treatment. In the undoped case considered here it can be shown that both mean-field schemes are equivalent due to particle-hole symmetry [4]. As already pointed out above, in the framework of Heisenberg systems, the flux and paired state order parameters are unphysical because they ignore the particle constraint. Hence, the corresponding broken symmetry is not a symmetry of the physical Heisenberg Hamiltonian but rather a broken gauge invariance of the auxiliary fermions. The respective phase transitions have to be interpreted as crossovers. In this thesis we will not consider mean-field amplitudes violating the auxiliary particle constraint.

By contrast, the magnetic order parameter $\langle \mathbf{S}_i \rangle = \frac{1}{2} \sum_{\alpha\beta} \langle f_{i\alpha}^\dagger \boldsymbol{\sigma}_{\alpha\beta} f_{i\beta} \rangle$ that appears in the Hartree approximation is a physical quantity. A finite magnetization corresponds to a real broken symmetry (the SU(2) symmetry of the original Heisenberg Hamiltonian) and the onset of magnetic order indicates a true phase transition.

Before we start with the calculations we summarize the Feynman rules needed in the following:

- Bare fermion propagators are given by

$$G^0_{ij,\alpha\beta}(i\omega) = \frac{1}{i\omega + \mu} \delta_{ij} \delta_{\alpha\beta} \quad \text{with} \quad \mu = \begin{cases} 0 & \text{average projection} \\ -\frac{i\pi}{2\beta} & \text{Popov-Fedotov method} \end{cases}.$$

4 Mean-Field Theory

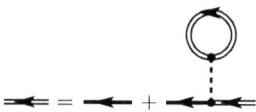

Figure 4.1: Diagrammatic representation of the Hartree approximation. The full line is the bare Green's function G^0, Eq. (3.6), the double-stroke line is the self consistent one. The dashed line represents the interaction J_1 or J_2 and the dots are Pauli matrices×1/2.

In diagrams they are represented by full lines with an arrow.

- Exchange interactions between the sites i and j contribute a factor $-J_{ij}$ and are illustrated by dashed lines.

- Bare vertices which are depicted by dots correspond to a factor $\frac{1}{2}\sigma^\mu_{\alpha\beta}$.

- Closed fermion lines carry a factor -1.

- There are sums over all internal indices, i.e., all Matsubara frequencies (fermionic $\frac{1}{\beta}\sum_{i\omega}\cdots$ or bosonic $\frac{1}{\beta}\sum_{i\nu}\cdots$), site indices i, spin variables α and spatial coordinates μ. A closed loop requires a trace in spin space.

Sums over Matsubara frequencies are evaluated using the theorem of residues,

$$\frac{1}{\beta}\sum_{i\omega} F(i\omega) = \sum_n \operatorname*{Res}_{z=z_n}(F(z)f(z)), \qquad (4.1a)$$

$$\frac{1}{\beta}\sum_{i\nu} B(i\nu) = -\sum_n \operatorname*{Res}_{z=z_n}(B(z)g(z)). \qquad (4.1b)$$

Here $f(z) = \frac{1}{e^{\beta z}+1}$ and $g(z) = \frac{1}{e^{\beta z}-1}$ denote the Fermi- and Bosefunction, respectively. z_n are the poles of the functions $F(z)$ or $B(z)$. We use the convention of writing fermionic Matsubara frequencies as $i\omega$ and bosonic Matsubara frequencies as $i\nu$.

In order to calculate ground-state properties it turns out to be convenient first to consider finite temperatures and performing the limit $T \equiv \frac{1}{\beta} \to 0$ in the end. Dyson's equation in Fig. 4.1 reads

$$\bar{G}_i(i\omega) = [(i\omega + \mu)\mathbb{1} - \bar{\Sigma}_i(i\omega)]^{-1}, \qquad (4.2)$$

where $\bar{G}$, $\bar{\Sigma}$ and $\mathbb{1}$ are matrices in spin space. The self energy is coupled back to the renormalized Green's function by

$$\bar{\Sigma}_i(i\omega) = \frac{1}{4}\sum_j J_{ij} \sum_{\mu=1}^{3} \sigma^\mu \frac{1}{\beta}\sum_{i\omega'} \operatorname{Tr}[\sigma^\mu \bar{G}_j(i\omega')]e^{i\omega'\delta}. \qquad (4.3)$$

The coupling J_{ij} equals J_1 if i,j are nearest neighbors, and J_2 if i,j are next-nearest neighbors. The factor $e^{i\omega\delta}$, with an infinitesimal $\delta > 0$, is needed for the convergence of

4 Mean-Field Theory

the Matsubara sum. If we assume magnetism along the z-direction, the self energy has the form

$$\bar{\Sigma}_i(i\omega) = \sigma^z m_i. \tag{4.4}$$

To describe Néel- and collinear order we split the lattice up into two sublattices A and B. In case of Néel order A and B form a staggered pattern while for collinear order they form rows (or equivalently columns). Furthermore we require

$$m \equiv m_{i \in A} = -m_{i \in B}. \tag{4.5}$$

Inserting Eq. (4.2) into Eq. (4.3) and using Eq. (4.4) one obtains

$$m_i = \frac{1}{4} \sum_j J_{ij} \frac{1}{\beta} \sum_{i\omega} \sum_{\zeta = \pm 1} \frac{\zeta}{i\omega + \mu - \zeta m_j} e^{i\omega \delta}. \tag{4.6}$$

Using $\frac{1}{\beta} \sum_{i\omega} \frac{e^{i\omega \delta}}{i\omega - z} = f(z)$ and $f(z - \mu^{\text{ppv}}) = \frac{1}{ie^{\beta z} + 1}$ (f is the Fermi function) one finds the following self-consistent equations for m for both types of order and both projection schemes,

$$\text{Néel-order: } m = \begin{cases} (J_1 - J_2)\tanh(\frac{m\beta}{2}) & \text{for } \mu = 0 \\ (J_1 - J_2)\tanh(m\beta) & \text{for } \mu = \mu^{\text{ppv}} \end{cases}, \tag{4.7a}$$

$$\text{collinear-order: } m = \begin{cases} J_2 \tanh(\frac{m\beta}{2}) & \text{for } \mu = 0 \\ J_2 \tanh(m\beta) & \text{for } \mu = \mu^{\text{ppv}} \end{cases}. \tag{4.7b}$$

The spin polarization or, in short, magnetization M_i is given by the bubble of a single renormalized propagator,

$$M_i = \langle S_i^z \rangle = \frac{1}{2} \frac{1}{\beta} \sum_{i\omega} \text{Tr}[\sigma^z \bar{G}_i(i\omega)] e^{i\omega \delta}. \tag{4.8}$$

From the comparison of Eq. (4.3) and Eq. (4.8) and using $M_{i \in A} = -M_{i \in B}$ one finds a relation between m_i and M_i,

$$m_i = \frac{1}{2} \sum_j J_{ij} M_j = \begin{cases} 2M_i(J_2 - J_1) & \text{for Néel-order} \\ -2M_i J_2 & \text{for collinear order} \end{cases}. \tag{4.9}$$

From Eqs. (4.7a) and (4.7b) the critical temperatures $T_c^{\text{Néel}}$ and T_c^{Col} can be determined. The instability with the larger transition temperature controls the type of order at a given $g = \frac{J_2}{J_1}$. This leads to (see also Fig. 4.2)

$$0 \leq g \leq \frac{1}{2} : T_c = T_c^{\text{Néel}} = \begin{cases} \frac{J_1}{2}(1-g) & \text{for } \mu = 0 \\ J_1(1-g) & \text{for } \mu = \mu^{\text{ppv}} \end{cases}, \tag{4.10a}$$

$$g \geq \frac{1}{2} : T_c = T_c^{\text{Col}} = \begin{cases} \frac{J_1}{2}g & \text{for } \mu = 0 \\ J_1 g & \text{for } \mu = \mu^{\text{ppv}} \end{cases}. \tag{4.10b}$$

4 Mean-Field Theory

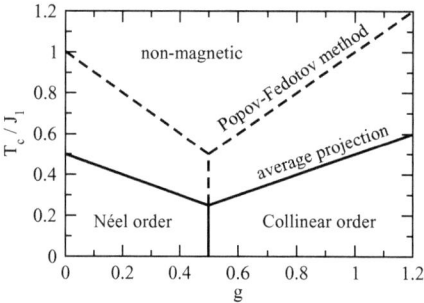

Figure 4.2: Phase diagram in the g-T-plane using the Hartree-approximation. The full lines are the phase boundaries with $\mu = 0$, the dashed lines use $\mu = -\frac{i\pi}{2\beta}$.

Apparently, within this approximation, no non-magnetic phase is found at $T = 0$. Instead there is a first order transition from Néel to collinear order at $g = \frac{1}{2}$. For $\beta \to \infty$ the Eqs. (4.7a) and (4.7b) lead to $|m| = J_1 - J_2$ ($|m| = J_2$) in the Néel case (collinear case). The magnetization $M = |M_i|$ which can now be obtained from Eq. (4.9) obviously reaches the saturation value $M = \frac{1}{2}$ at $T = 0$, and the classical large spin behavior is reproduced. This was expected since in a mean-field treatment quantum fluctuations are neglected such that the system behaves classically. While at $T = 0$ the magnetization is the same for both projection schemes, this is no longer the case for $T > 0$. The contribution of unphysical states with $S = 0$ leads to a reduction of the magnetization in the average projection scheme. Also the critical temperatures come out by a factor of two smaller, see Fig. 4.2. The self-consistent equations for $\mu = \mu^{\mathrm{ppv}}$ are identical to those obtained within the conventional mean-field theory in terms of spin operators, confirming that the cancellation of the unphysical states works correctly in this approximation.

In summary, the simple mean-field theory leads to a Néel phase at $g < \frac{1}{2}$ and a collinear-ordered phase $g > \frac{1}{2}$ but is insufficient to describe the effect of frustration in destroying magnetic order in the regime $g \approx \frac{1}{2}$. Nevertheless, we stick a bit longer to the mean-field scheme to learn about the dynamical properties of magnetically ordered phases.

4.2 Random-Phase Approximation

The Hartree scheme applied in the previous section computes the magnetization as a natural outcome of the self-consistency equations. In order to calculate response functions such as the susceptibility or excitation spectra within a mean-field treatment, we now employ the random-phase approximation (RPA). Fig. 4.3 displays the approximation in diagrammatic form. Since the RPA scheme can be obtained from the Hartree approximation by taking the derivative with respect to the self consistent field, phase transitions are located at the same point in both approaches. An important aspect of such a scheme is that RPA in conjunction with the Hartree self-energy is a conserving

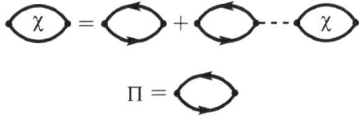

Figure 4.3: Self-consistent RPA equation for the susceptibility χ in diagrammatic representation. Π denotes a single bubble.

approximation in the sense of Baym and Kadanoff. We emphasize that the auxiliary-particle constraint is meaningful only if particle number conservation is guaranteed. We now compute the RPA susceptibility χ in order to find the excitation spectrum at $T = 0$. The RPA equation in real space reads

$$\chi_{ij}^{\mu\mu'} = \Pi_{ij}^{\mu\mu'} - (\Pi J \chi)_{ij}^{\mu\mu'}, \qquad (4.11)$$

with Π being the fermion bubble shown in Fig. 4.3. We note that the propagators that enter Π are those containing the Hartree self-energy, i.e., the renormalized propagators from Fig. 4.1. Of course $\Pi_{ij}^{\mu\mu'} = \Pi_i^{\mu\mu'} \delta_{ij}$ is a local quantity but in non-uniform magnetic phases the Hartree self-energy still leads to a dependence on the sublattice $A, B \ni i$. Applying the Feynman rules, Π has the form

$$\Pi_i^{\mu\mu'}(i\nu) = -\frac{1}{4}\frac{1}{\beta}\sum_{i\omega}\text{Tr}[\sigma^\mu \bar{G}_i(i\omega + i\nu)\sigma^{\mu'}\bar{G}_i(i\omega)]. \qquad (4.12)$$

Inserting Eq. (4.2) with Eq. (4.4) into Eq. (4.12) and performing the Matsubara sum and the trace yields the expressions

$$\Pi_i^{xx}(i\nu) = \Pi_i^{yy}(i\nu) = \frac{m_i}{(2m_i)^2 - (i\nu)^2} \times \begin{cases} \tanh\left(\frac{\beta m_i}{2}\right) & \text{for } \mu = 0 \\ \tanh(\beta m_i) & \text{for } \mu = \mu^{\text{ppv}} \end{cases}, \qquad (4.13a)$$

$$\Pi_i^{xy}(i\nu) = -\Pi_i^{yx}(i\nu) = \frac{1}{2i}\frac{i\nu}{(2m_i)^2 - (i\nu)^2} \times \begin{cases} \tanh\left(\frac{\beta m_i}{2}\right) & \text{for } \mu = 0 \\ \tanh(\beta m_i) & \text{for } \mu = \mu^{\text{ppv}} \end{cases}, \qquad (4.13b)$$

and $\Pi_B^{xx} = \Pi_A^{xx}$, $\Pi_B^{xy} = -\Pi_A^{xy}$. For brevity we omit the frequency arguments in the following. All components with indices "xz" or "yz" vanish. Moreover, Π^{zz} and χ^{zz} are of no relevance here because in the limit $T \to 0$ where all spins are maximally polarized, the longitudinal response decreases exponentially. Eq. (4.11) can be solved by Fourier transform. Due to the spatial dependence of Π_i the susceptibility χ_{ij} is not just a function of the distance vector $\mathbf{R}_i - \mathbf{R}_j$. The Fourier transform of Π_{ij} and χ_{ij} hence involves two wave vectors,

$$\Pi_{ij}^\mu = \frac{1}{N^2}\sum_{\mathbf{k}\mathbf{k}'} e^{i(\mathbf{k}\mathbf{R}_i - \mathbf{k}'\mathbf{R}_j)} \Pi^{\mu\mu'}(\mathbf{k},\mathbf{k}'), \qquad (4.14a)$$

$$\chi_{ij}^\mu = \frac{1}{N^2}\sum_{\mathbf{k}\mathbf{k}'} e^{i(\mathbf{k}\mathbf{R}_i - \mathbf{k}'\mathbf{R}_j)} \chi^{\mu\mu'}(\mathbf{k},\mathbf{k}'), \qquad (4.14b)$$

while $J_{ij} = J(\mathbf{R}_i - \mathbf{R}_j)$ is a function of the distance and hence

$$J_{ij} = \frac{1}{N} \sum_{\mathbf{k}} e^{i\mathbf{k}(\mathbf{R}_i - \mathbf{R}_j)} J(\mathbf{k}) \qquad (4.15)$$

with

$$J(\mathbf{k}) = 2J_1[\cos(k_x) + \cos(k_y)] + 4J_2 \cos(k_x)\cos(k_y). \qquad (4.16)$$

Transforming Eq. (4.11) into Fourier space we obtain

$$\chi^{\mu\mu'}(\mathbf{k},\mathbf{k}') = \Pi^{\mu\mu'}(\mathbf{k},\mathbf{k}') - \frac{1}{N}\sum_{\mathbf{q}}(\Pi(\mathbf{k},\mathbf{q})J(\mathbf{q})\chi(\mathbf{q},\mathbf{k}'))^{\mu\mu'}. \qquad (4.17)$$

The diagonal components Π_i^{xx} and Π_i^{yy} do not depend on the sublattice whereas the off-diagonal components Π_i^{xy} and Π_i^{yx} have opposite sign on sublattices A and B. Thus the transverse subspace of $\Pi^{\mu\mu'}(\mathbf{k},\mathbf{k}')$ has the form

$$\Pi^{\mu\mu'}(\mathbf{k},\mathbf{k}')\bigg|_{\mu,\mu'=x,y} = N \begin{pmatrix} \Pi_A^{xx} \delta_{\mathbf{k},\mathbf{k}'} & \Pi_A^{xy} \delta_{\mathbf{k},\mathbf{k}'\pm\mathbf{Q}} \\ -\Pi_A^{xy} \delta_{\mathbf{k},\mathbf{k}'\pm\mathbf{Q}} & \Pi_A^{xx} \delta_{\mathbf{k},\mathbf{k}'} \end{pmatrix}, \qquad (4.18)$$

with the ordering vector $\mathbf{Q}$ given by $\mathbf{Q}_{\text{Néel}} = (\pi,\pi)$ in the Néel phase and $\mathbf{Q}_{\text{Col,1}} = (\pi,0)$ or $\mathbf{Q}_{\text{Col,2}} = (0,\pi)$ in the collinear phase. Furthermore we assume $e^{\pm i\mathbf{Q}\mathbf{R}_i} = +1,-1$ for $\mathbf{R}_i \in A,B$, respectively. Multiplying two of such bubbles yields

$$\frac{1}{N}\sum_{\mathbf{q}} N \begin{pmatrix} \Pi_A^{xx} \delta_{\mathbf{k},\mathbf{q}} & \Pi_A^{xy} \delta_{\mathbf{k},\mathbf{q}\pm\mathbf{Q}} \\ -\Pi_A^{xy} \delta_{\mathbf{k},\mathbf{q}\pm\mathbf{Q}} & \Pi_A^{xx} \delta_{\mathbf{k},\mathbf{q}} \end{pmatrix} N \begin{pmatrix} \Pi_A^{xx} \delta_{\mathbf{q},\mathbf{k}'} & \Pi_A^{xy} \delta_{\mathbf{q},\mathbf{k}'\pm\mathbf{Q}} \\ -\Pi_A^{xy} \delta_{\mathbf{q},\mathbf{k}'\pm\mathbf{Q}} & \Pi_A^{xx} \delta_{\mathbf{q},\mathbf{k}'} \end{pmatrix}$$

$$= N \begin{pmatrix} [(\Pi_A^{xx})^2 - (\Pi_A^{xy})^2]\delta_{\mathbf{k},\mathbf{k}'} & 2\Pi_A^{xx}\Pi_A^{xy}\delta_{\mathbf{k},\mathbf{k}'\pm\mathbf{Q}} \\ -2\Pi_A^{xx}\Pi_A^{xy}\delta_{\mathbf{k},\mathbf{k}'\pm\mathbf{Q}} & [(\Pi_A^{xx})^2 - (\Pi_A^{xy})^2]\delta_{\mathbf{k},\mathbf{k}'} \end{pmatrix} \qquad (4.19)$$

with the same form as a single bubble. Hence, building up the RPA bubble chain, the momentum dependence of Π is retained and the susceptibility χ can be written as

$$\chi^{\mu\mu'}(\mathbf{k},\mathbf{k}')\bigg|_{\mu,\mu'=x,y} = N \begin{pmatrix} \chi^{xx}(\mathbf{k})\delta_{\mathbf{k},\mathbf{k}'} & \chi^{xy}(\mathbf{k})\delta_{\mathbf{k},\mathbf{k}'\pm\mathbf{Q}} \\ -\chi^{xy}(\mathbf{k})\delta_{\mathbf{k},\mathbf{k}'\pm\mathbf{Q}} & \chi^{xx}(\mathbf{k})\delta_{\mathbf{k},\mathbf{k}'} \end{pmatrix}. \qquad (4.20)$$

Indeed, this ansatz together with Eq. (4.18) solves Eq. (4.17) yielding the matrix relation

$$\begin{pmatrix} 1 + \Pi_A^{xx}J(\mathbf{k}) & -\Pi_A^{xy}J(\mathbf{k}\pm\mathbf{Q}) \\ \Pi_A^{xy}J(\mathbf{k}) & 1 + \Pi_A^{xx}J(\mathbf{k}\pm\mathbf{Q}) \end{pmatrix} \begin{pmatrix} \chi^{xx}(\mathbf{k}) \\ \chi^{xy}(\mathbf{k}\pm\mathbf{Q}) \end{pmatrix} = \begin{pmatrix} \Pi_A^{xx} \\ \Pi_A^{xy} \end{pmatrix}. \qquad (4.21)$$

Inverting the matrix we obtain an expression for the transverse susceptibility $\chi^{xx}(\mathbf{k})$,

$$\chi^{xx}(\mathbf{k}) = \frac{\Pi_A^{xx} + [(\Pi_A^{xx})^2 + (\Pi_A^{xy})^2]J(\mathbf{k}\pm\mathbf{Q})}{(1 + \Pi_A^{xx}J(\mathbf{k}))(1 + \Pi_A^{xx}J(\mathbf{k}\pm\mathbf{Q})) + (\Pi_A^{xy})^2 J(\mathbf{k})J(\mathbf{k}\pm\mathbf{Q})}. \qquad (4.22)$$

4 Mean-Field Theory

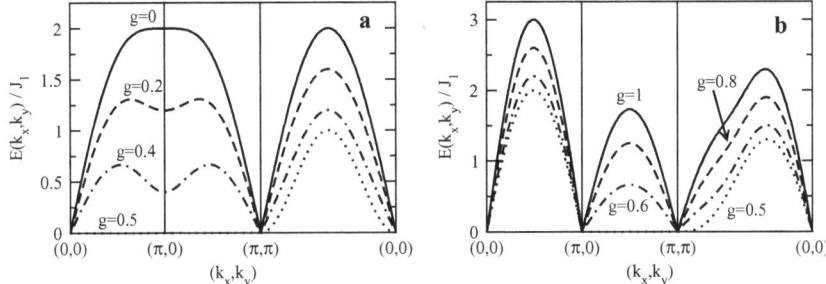

Figure 4.4: Dispersion of the spin excitations in the Néel phase (a) and in the collinear phase (b) within an RPA scheme. Shown are the excitation energies for a path along lines of high symmetry in the Brillouin zone.

Finally we perform the limitation $\beta \to \infty$ in the Eqs. (4.13a) and (4.13b). We use Eq. (4.9) with $M = \frac{1}{2}$ and insert the resulting expressions for Π^{xx} and Π^{xy} into Eq. (4.22). It follows

$$\chi^{xx}_{\text{Néel}}(\mathbf{k}) = \frac{J_1 - J_2 + J(\mathbf{k} \pm \mathbf{Q}_{\text{Néel}})/4}{2E_{\text{Néel}}(\mathbf{k})} \left(\frac{1}{E_{\text{Néel}}(\mathbf{k}) - i\nu} + \frac{1}{E_{\text{Néel}}(\mathbf{k}) + i\nu} \right) \quad \text{with}$$

$$E_{\text{Néel}}(\mathbf{k}) = \sqrt{\frac{1}{4}(4J_1 + J(\mathbf{k}))(4J_1 + J(\mathbf{k} \pm \mathbf{Q}_{\text{Néel}})) - J_2(8J_1 - 4J_2 + J(\mathbf{k}) + J(\mathbf{k} \pm \mathbf{Q}_{\text{Néel}}))}$$

(4.23)

in the case of Néel order and

$$\chi^{xx}_{\text{Col},1}(\mathbf{k}) = \frac{J_2 + J(\mathbf{k} \pm \mathbf{Q}_{\text{Col},1})/4}{2E_{\text{Col},1}(\mathbf{k})} \left(\frac{1}{E_{\text{Col},1}(\mathbf{k}) - i\nu} + \frac{1}{E_{\text{Col},1}(\mathbf{k}) + i\nu} \right) \quad \text{with}$$

$$E_{\text{Col},1}(\mathbf{k}) = \frac{1}{2}\sqrt{(4J_2 + J(\mathbf{k}))(4J_2 + J(\mathbf{k} \pm \mathbf{Q}_{\text{Col},1}))},$$

(4.24)

for collinear order (with ordering vector $\mathbf{Q}_{\text{Col},1} = (\pi, 0)$). In order to obtain a physical quantity we perform an analytic continuation to the real axis simply by replacing $i\nu \to \omega + i\delta$. The magnon spectral-function is then given by $\frac{1}{\pi}\text{Im}\chi^{xx}(\mathbf{k}, \omega + i\delta)$. The identity $\text{Im}\frac{1}{E-\omega\pm i\delta} = \mp\delta(\omega - E)$ converts the terms in the brackets of Eqs. (4.23) and (4.24) into δ-functions indicating excitations in the form of undamped spin waves. Their dispersion is simply given by

$$\omega_{\text{Néel/Col},1} = E_{\text{Néel/Col},1}(\mathbf{k}).$$

(4.25)

Fig. 4.4 shows the dispersion of the spin waves in the Néel phase, $0 \leq g \leq 0.5$, and in the collinear phase, $0.5 \leq g \leq 1$. In case of Néel order, see Fig. 4.4a, two wave vectors with vanishing excitation energy can be identified, $\mathbf{k} = (0,0)$ and $\mathbf{k} = (\pi,\pi)$, where the latter is the Goldstone mode. According to the Goldstone theorem broken SU(2) symmetry in the form of long-range magnetic order must be accompanied with such a mode at the respective ordering vector Q. In contrast, the mode with $\mathbf{k} = (0,0)$ is no Goldstone mode, which can be seen from the first line of Eq. (4.23) where the spectral weight given by the factor outside the bracket vanishes at $\mathbf{k} = (0,0)$. In general, the excitation energies and the spin-wave velocity decrease with increasing frustration g until in the limit of maximal frustration $g = 0.5$ the excitation energy vanishes along the path $\mathbf{k} = (0,0) \rightarrow (\pi,0) \rightarrow (\pi,\pi)$ indicating the large degeneracy in the classical model.

Evidence for the first order transition at $g = 0.5$ arises from a comparison to Fig. 4.4b where the collinear phase is studied. The dispersion changes discontinuously once the magnetic order switches from Néel to collinear. Now the Goldstone mode resides at the ordering vector $\mathbf{Q} = (\pi,0)$ while other wave vectors with vanishing excitation energy, i.e., $\mathbf{k} = (0,0)$ and $\mathbf{k} = (\pi,\pi)$, have no spectral weight. With increasing g the degeneracy at $g = 0.5$ on the line between $\mathbf{k} = (\pi,0)$ and $\mathbf{k} = (\pi,\pi)$ is lifted and the spin-wave velocity rises.

These results demonstrate that even on a mean-field level magnetic order is described qualitatively correct as long as a single kind of magnetic order is considered. On the other hand, frustration effects such as the melting of long range order induced by competing interactions are beyond the scope of a simple mean-field treatment. In order to overcome this problem we extend the mean-field scheme in the next chapter.

5 Finite Pseudo-Fermion Lifetime

In the mean-field approximation of the previous chapter, the effect of fermion scattering in generating a finite lifetime of the pseudo fermions has not been taken into account. The most natural way to include such effects is to approximate the self energy by some set of diagrammatic contributions. However, as will be discussed in Section 5.3, finding a reasonable approximative scheme is a non-trivial issue. This can be understood if one realizes that a small expansion parameter is absent in our Hamiltonian. Hence, approximations require infinite resummations of perturbation theory in the couplings J_1 and J_2. In Sections 5.1 and 5.2 we put forward a different and somehow simpler route. Instead of calculating the self energy Σ we make an ansatz that corresponds to an exponential decay of the propagator with the pseudo-fermion lifetime τ as a phenomenological parameter. Of course there is still the problem of finding a numerical value for this parameter τ, however, it will turn out that even on a phenomenological level this scheme captures the essential aspects of competing spin interactions.

We model the retarded Green's function by,

$$G^{\mathrm{R}}(\omega) = \frac{1}{\omega + i\gamma}, \quad \Sigma = -i\gamma \quad \text{with} \quad \gamma = \frac{1}{\tau}. \tag{5.1}$$

In real-time representation this propagator exhibits an exponential decay,

$$G^{\mathrm{R}}(t - t') = \frac{\Theta(t - t')}{i} e^{-\frac{t-t'}{\tau}}. \tag{5.2}$$

The spectral function $\rho(\omega)$ has the form of a Lorentzian with the (finite) width γ,

$$\rho(\omega) = -\frac{1}{\pi} \operatorname{Im} G^{\mathrm{R}}(\omega) = \frac{\gamma}{\pi} \frac{1}{\omega^2 + \gamma^2}. \tag{5.3}$$

It proved to be convenient to perform calculations on the imaginary Matsubara axis. According to Eq. (5.1) the self energy Σ is constant for frequencies slightly above the real axis. Thus the analytic continuation to the upper complex half plane is trivial and provides

$$\Sigma(z) = -i\gamma \quad \text{for} \quad \operatorname{Im} z > 0. \tag{5.4}$$

A second statement about the self energy on the Matsubara axis can be made regarding the spectral representation $G(i\omega) = \int_{-\infty}^{\infty} \frac{\rho(\epsilon)}{i\omega - \epsilon} d\epsilon$. Due to particle-hole symmetry, $\rho(\omega)$ must be an even function. It follows immediately that $G(i\omega)$ is an odd function with vanishing real part along the Matsubara axis. Because of Dyson's equation $G(i\omega) =$

$\frac{1}{i\omega - \Sigma(i\omega)}$ the same properties also hold for $\Sigma(i\omega)$. These statements completely determine $G(i\omega)$ and $\Sigma(i\omega)$,

$$G(i\omega) = \frac{1}{i\omega + i\gamma\,\mathrm{sgn}(\omega)}\,, \quad \Sigma(i\omega) = -i\gamma\,\mathrm{sgn}(\omega)\,. \tag{5.5}$$

The damping parameter γ has the dimension of an energy. To proceed, we need to specify its dependence on the couplings J_1 and J_2. For $J_2 = 0$, we put γ in the form $\gamma = \tilde{\gamma} J_1$, where $\tilde{\gamma}$ is a dimensionless parameter. A similar situation is encountered for $J_1 \to 0$ and $J_2 > 0$, where the system is split up into two square lattices, each only with nearest-neighbor couplings J_2. Therefore, in this limit, the relation $\gamma = \tilde{\gamma} J_2$ holds. To interpolate between both limiting cases, we assume

$$\gamma(J_1, J_2) = \tilde{\gamma} J_1 \sqrt{1 + g^2}\,. \tag{5.6}$$

Later we will see that the precise choice of the interpolation function is of minor relevance for the results.

5.1 Hartree Approximation

In order to calculate the ground-state magnetization we now repeat the Hartree approximation of Section 4.1 but with the bare Green's function replaced by Eq. (5.5). In the limit $T \to 0$, using $\frac{1}{\beta}\sum_{i\omega} \to \frac{1}{2\pi}\int d\omega$, Eq. (4.6) translates into the new mean-field equation given by

$$m_i = \frac{1}{4}\sum_j J_{ij} \frac{1}{2\pi}\int_{-\infty}^{\infty} d\omega \sum_{\zeta=\pm 1} \frac{\zeta}{i\omega + i\gamma\,\mathrm{sgn}(\omega) - \zeta m_j}\,. \tag{5.7}$$

Here it is obvious that the two projection schemes are identical because a shift of the Matsubara frequencies by $\mu^{\mathrm{ppv}} = -\frac{i\pi}{2\beta}$ becomes irrelevant in the limit $T \to 0$, provided that the Green's function, or equivalently the fermion spectral-function, is regular at $\omega = 0$. The integral in Eq. (5.7) is straightforwardly evaluated. Using Eq. (4.9) we obtain the following self-consistent equations for the Néel- and collinear magnetizations,

$$M_{\mathrm{Néel}} = \frac{1}{\pi}\arctan\left[\frac{2M_{\mathrm{Néel}}(J_1 - J_2)}{\gamma}\right]\,, \tag{5.8a}$$

$$M_{\mathrm{Col}} = \frac{1}{\pi}\arctan\left(\frac{2M_{\mathrm{Col}}J_2}{\gamma}\right)\,. \tag{5.8b}$$

The solutions of these equations (using the interpolation in Eq. (5.6)) are shown in Fig. 5.1 for different parameters $\tilde{\gamma}$. The case $\tilde{\gamma} = 0$ represents the Hartree approximation from Section 4.1. An increase in $\tilde{\gamma}$ reduces the magnetizations, especially in the region of high frustration. In particular, for small $\tilde{\gamma}$, there is still a direct first-order transition between the two types of order at $g = \frac{1}{2}$, while for sufficiently large $\tilde{\gamma}$ a non-magnetic

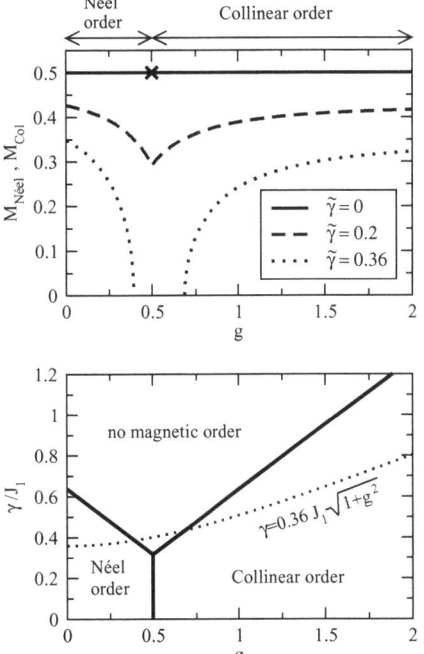

Figure 5.1: Magnetizations $M_{\text{Néel}}$ and M_{Col} versus g within a Hartree approximation assuming a finite pseudo-fermion lifetime τ. The dependence of $\gamma = \frac{1}{\tau}$ on the couplings J_1 and J_2 is chosen according to Eq. (5.6). Shown are different values for $\tilde{\gamma}$.

Figure 5.2: Phase diagram in the γ-g-plane. The dotted line shows the g-dependence of γ according to Eq. (5.6) for $\tilde{\gamma} = 0.36$.

phase emerges. It appears that a broadening of the pseudo-fermion levels captures much of the effect of frustration expected to reduce or destroy magnetic order. In contrast to the simple mean-field theory, one now finds second order phase transitions and a mean-field critical exponent $\beta = \frac{1}{2}$ of the magnetization. From the self-consistent equations, a phase diagram in the γ-g-plane can be drawn, see Fig. 5.2. This diagram demonstrates that the form of the interpolation function $\gamma(J_1, J_2)$ has an effect mainly on how symmetric the non-magnetic phase is located around $g = 0.5$. Additionally, it shows only a narrow parameter range for γ where the theory provides meaningful values for the phase boundaries. This illustrates that it will be difficult to determine the damping parameter in approximative schemes. For example $\tilde{\gamma} = 0.36$ leads to transitions at $g_{c1} \approx 0.39$, $g_{c2} \approx 0.69$ and also a realistic value for the magnetization at $g = 0$, i.e., $M_{\text{Néel}} \approx 0.35$. This value of the width parameter $\tilde{\gamma}$ will be used in the following section to study the properties of the non-magnetic phase.

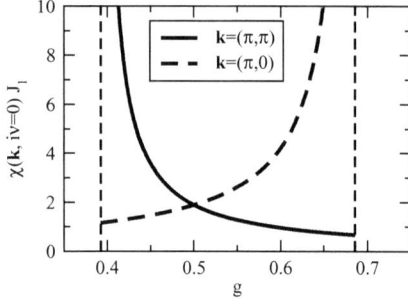

Figure 5.3: Static susceptibility for wave vectors (π,π) and $(\pi,0)$ within an RPA scheme employing a finite damping parameter $\tilde{\gamma} = 0.36$. The thin dashed lines visualize the phase boundaries.

5.2 Random-Phase Approximation

With the scheme introduced in the previous section we have a method at hand that generates a non-magnetic phase, provided that the damping γ is adjusted correctly. It has been shown that this parameter controls the extent of the ground-state phases. This gives us the opportunity to calculate the properties of the paramagnetic phase in more detail. A suitable quantity to study is the spin susceptibility as it also leads to related observables such as the dynamic structure factor, the correlation function and the correlation length. Our approach is similar to Section 4.2. We employ the RPA scheme, see Fig. 4.3, but use the Green's function introduced in Eq. (5.5). As long as we consider the paramagnetic phase, SU(2) symmetry is conserved, i.e., $\Pi^{\mu\mu'} = \Pi\delta_{\mu\mu'}$, $\chi^{\mu\mu'} = \chi\delta_{\mu\mu'}$ and all quantities are translation invariant, $\Pi_{ij} = \Pi\delta_{ij}$, $\chi_{ij} = \chi(\mathbf{R}_i - \mathbf{R}_j)$. This simplifies the calculation of Section 4.2 considerably because the Fourier transform of Eq. (4.11) can be performed straightforwardly, yielding

$$\chi(\mathbf{k},i\nu) = \frac{1}{[\Pi(i\nu)]^{-1} + J(\mathbf{k})}. \tag{5.9}$$

First we shall calculate the static susceptibility. We need the bubble $\Pi(i\nu = 0)$ using the propagator from Eq. (5.5). This quantity is found as

$$\Pi(i\nu = 0) \equiv \Pi^{zz}(i\nu=0) = -\frac{1}{4}\frac{1}{2\pi}\int d\omega \left(\frac{1}{i\omega + i\gamma\,\text{sgn}(\omega)}\right)^2 \text{Tr}\left[(\sigma^z)^2\right] = \frac{1}{2\pi\gamma}. \tag{5.10}$$

The susceptibility in Eq. (5.9) together with Eqs. (4.16), (5.6) and (5.10) is evaluated for $\mathbf{k} = (\pi,\pi)$ and $\mathbf{k} = (\pi,0)$, the relevant wave vectors in the case of Néel- and collinear order, respectively. The results for $\tilde{\gamma} = 0.36$ are shown in Fig. 5.3. As expected for continuous phase transitions, the susceptibility with wave vector $\mathbf{k} = (\pi,\pi)$ ($\mathbf{k} = (0,\pi)$) diverges in the limit $g_c \to g_{c1} + 0$ ($g_c \to g_{c2} - 0$).

Dynamical properties of the paramagnetic phase may also be obtained within an analytical treatment. For the dynamical susceptibility $\chi(\mathbf{k},i\nu)$ we need Eq. (5.10) at finite frequencies. Evaluating the integral, we obtain

$$\Pi(i\nu) = -\frac{1}{4}\frac{1}{2\pi}\int dw \frac{1}{[i\omega + i\nu + i\gamma\,\text{sgn}(\omega+\nu)][i\omega + i\gamma\,\text{sgn}(\omega)]}\text{Tr}\left[(\sigma^z)^2\right]$$
$$= \frac{\gamma \ln\left(1+\frac{\nu}{\gamma}\right)}{\pi\nu(2\gamma+\nu)}. \tag{5.11}$$

Inserting this quantity into Eq. (5.9) we gain an expression for the dynamical susceptibility. The analytic continuation is performed by replacing $\nu \to -i\omega$. We note that an infinitesimal damping $\delta \ll J_1, J_2$ is not needed in this replacement because it becomes irrelevant compared to the large damping $\gamma \approx J_1, J_2$.

The dynamical spin-structure factor

$$S(\mathbf{k},\omega) = \frac{1}{\pi}\frac{1}{1-e^{-\frac{\omega}{T}}}\text{Im}\,\chi(\mathbf{k},\omega) \tag{5.12}$$

is an important quantity since it is the canonical outcome of neutron-scattering experiments and a measure for the dynamics of spin fluctuations. At zero temperature the factor $\frac{1}{1-e^{-\frac{\omega}{T}}}$ is unity for $\omega > 0$ and zero otherwise. Evaluating Eq. (5.12) we end up with

$$S(\mathbf{k},\omega) = \frac{\gamma\omega\left[\omega \arctan(\frac{\omega}{\gamma}) + \gamma \ln\left(1+(\frac{\omega}{\gamma})^2\right)\right]}{\left[\frac{1}{2}J(\mathbf{k})\gamma \ln\left(1+(\frac{\omega}{\gamma})^2\right) - \pi\omega^2\right]^2 + \left[J(\mathbf{k})\gamma\arctan(\frac{\omega}{\gamma}) + 2\pi\gamma\omega\right]^2} \tag{5.13}$$

In order to illustrate this function we plot it for different frustrations g and momenta $\mathbf{k}$ on the line from $\mathbf{k} = (\pi,\pi)$ to $\mathbf{k} = (\pi,0)$, see Fig. 5.4. At the transition point between the Néel- and the paramagnetic phase, i.e., at $g = g_{c1}$ the emerging Goldstone mode manifests itself as a divergence of the (π,π) component at $\omega = 0$. For momenta away from the Néel ordering vector this divergence is regularized and peaks at finite frequency appear, indicating damped spinwaves. Approaching the collinear ordering vector $\mathbf{k} = (\pi,0)$ the maxima shift towards higher frequencies, the height decreases and a larger width is acquired. The peak position gives an indication of the dispersion of the damped magnons. When we go further inside the paramagnetic phase, e.g. to $g = 0.45$, the divergence of the (π,π) component disappears. Moreover, the dispersion becomes flatter. This trend continues for an approach to $g = 0.5$ where the dispersion is completely flat along the line of considered momenta. Here, all curves lie exactly on top of each other. Hence, it demonstrates the high degeneracy of low lying excitations at large frustration. Obviously, the point $g = 0.5$ has still a unique character in our approximation scheme. Increasing g beyond $g = 0.5$ collinear fluctuations become the dominant ones. In this frustration regime the spin dynamics resemble those below $g = 0.5$ but with interchanged roles of $\mathbf{k} = (\pi,\pi)$ and $\mathbf{k} = (\pi,0)$. Finally at $g = g_{c2}$ the divergence of the $(\pi,0)$ component indicates the Goldstone mode of the collinear ordered phase. Throughout the paramagnetic phase the peak corresponding to the dominant

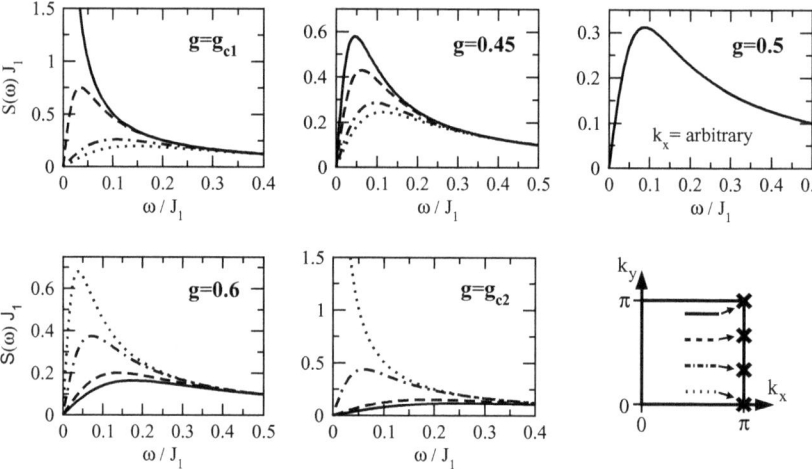

Figure 5.4: Dynamical spin structure factor in the paramagnetic phase for various frustrations g. The full, dashed, dashed-dotted and dotted curve represent momenta on the line from $\mathbf{k} = (\pi, \pi)$ to $\mathbf{k} = (\pi, 0)$ in steps of $\frac{\pi}{3}$, respectively (see the picture in the lower right corner).

spin-wave momentum resides at a frequency smaller than $0.1J_1$, representing a rather small energy scale.

Finally, we discuss the static correlation function $\chi(\mathbf{R}, i\nu = 0)$, which is obtained by transforming the susceptibility from Eq. (5.9) into real space,

$$\chi(\mathbf{R}, i\nu = 0) = \frac{1}{(2\pi)^2} \int_{-\pi}^{\pi} dk_x \int_{-\pi}^{\pi} dk_y \frac{e^{i\mathbf{k}\mathbf{R}}}{[\Pi(i\nu = 0)]^{-1} + J(\mathbf{k})} . \quad (5.14)$$

Evaluating Eq. (5.14) numerically with $\tilde{\gamma} = 0.36$ for distances R along a lattice direction, $\mathbf{R}_\mu = R\mathbf{e}_\mu$, $\mu = x, y$, leads to the behavior shown in Fig. 5.5. For g slightly above the lower critical value g_{c1} (upper panel) the signature of the Néel phase is clearly seen. The correlation function forms a staggered pattern and the envelopes for positive and negative data points only differ by a sign. At large enough distances R the correlations are well fitted by an exponential decay while at small distances the decrease is faster. Inside the paramagnetic phase the envelopes are no longer symmetric around $\chi(R) = 0$. For g slightly below the upper critical point g_{c2} (lower panel) the correlation function still exhibits a staggered sign but the correlation between spins with an odd distance seems to vanish on approaching the critical point. Again, for large R an exponential function can be fitted to both branches and the correlation length is identical for even and odd distances. The asymmetry of the two envelopes can be understood by the fact that for

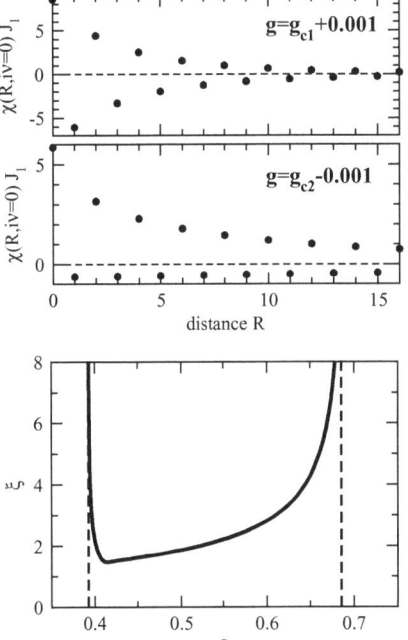

Figure 5.5: Static correlation function $\chi(R, i\nu = 0)$ for distances $R = |\mathbf{R}|$ along a lattice direction. R is measured in units of a lattice spacing. Again $\tilde{\gamma} = 0.36$ is used. In the upper panel g is slightly above the critical point g_{c1} and in the lower panel slightly below g_{c2}.

Figure 5.6: Static correlation length ξ in units of a lattice spacing versus g. The spectral width is $\tilde{\gamma} = 0.36$.

collinear fluctuations, two degenerate patterns exist, the alignment of spins along rows and along columns. Thus, near the upper critical point correlations are a superposition of both,

$$\chi(R) = (-1)^R a_1 e^{-\frac{R}{\xi}} + a_2 e^{-\frac{R}{\xi}} \qquad (5.15)$$

with (almost) identical weights $a_1 = a_2$. Obviously this suppresses correlations for odd distances. Here ξ denotes the correlation length. Away from the upper critical point Néel-like fluctuations emerge and we have $a_1 > a_2$. Eventually at the lower critical point a_2 vanishes.

The correlation length ξ is plotted in Fig. 5.6. The data indicate divergences at the phase boundaries but ξ gets rather small in the vicinity of $g = 0.4$, i.e., down to $\xi \approx 1.5$. Remarkably, the smallest values for the correlation length are not reached at $g = 0.5$ where one would classically expect the strongest frustration.

In conclusion, the phenomenological theory presented in this chapter suggests that a broadening of the fermions' spectral function controls the phase diagram and the behavior of many physical quantities like the magnetization, the susceptibility and the spatial correlation function. It leads to a qualitatively correct description of the paramagnetic phase in terms of these observables. However, a statement about the nature of the paramagnetic phase (columnar dimer or plaquette order) cannot be made. Also critical

5 Finite Pseudo-Fermion Lifetime

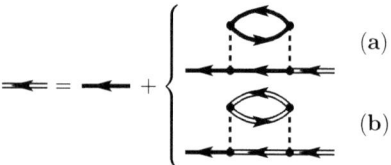

Figure 5.7: Dyson's equation for the renormalized propagator (double-stroke line) using the bare second-order self energy (a) or the completely renormalized second-order self energy (b). Full lines present bare propagators.

behavior beyond mean-field is not accessible. In order to obtain quantitatively correct transition points, the tuning parameter $\tilde{\gamma}$ has to be fixed to $\tilde{\gamma} \approx 0.36$. However, this phenomenological parameter cannot be determined within the theory and there is no simple way to calculate it. This problem will be briefly discussed in the next section.

5.3 The Spectral Width in Diagrammatic Approximations

We emphasize again that a small expansion parameter is absent in the models considered here because we operate in the strong coupling limit. Consequently there is no justification for a perturbative treatment in finite order. Instead diagram classes have to be summed up to infinite order in a self consistent way. A well elaborated resummation scheme is given by the concept of a conserving approximation already applied in Chapter 4 in the form of a RPA + Hartree scheme. Also the finite-lifetime approach in Sections 5.1 and 5.2 was a conserving approximation but with a redefinition of the bare propagator. However, from a numerical point of view it turns out to be very challenging to calculate a conserving scheme beyond RPA + Hartree. For this reason it is very difficult to gain a reasonable value for the damping parameter γ by summing up diagrammatic contributions of the Green's function [23, 24]. Away from conserving schemes this problem manifests itself in the fact that results might vary a lot depending on the choice of diagrams. To exemplify this behavior we have calculated the spectral function of the renormalized propagators in Fig. 5.7. In a first scheme we approximate the self energy by the second order term, see Fig. 5.7a, which is the lowest order of a non-vanishing contribution. (Note that the Hartree term is of first order in the couplings but it is finite only if $SU(2)$ symmetry breaking is assumed self consistently.) Calculating the spectral function of the renormalized propagator using the average projection with $\mu = 0$ one obtains

$$\rho(\omega) = \frac{1}{2}[\delta(\omega - \Delta) + \delta(\omega + \Delta)] \tag{5.16}$$

with

$$\Delta = \sqrt{\frac{3}{8}(J_1^2 + J_2^2)}. \tag{5.17}$$

Inserting the renormalized propagator of Fig. 4.1 into the Hartree scheme of Fig. 5.7a leads to a destruction of magnetic order in the whole parameter regime $g \geq 0$. This

5 Finite Pseudo-Fermion Lifetime

behavior can be traced back to the spectral width $\gamma \sim \Delta$ which apparently comes out much too large to allow for magnetic phases. On the other hand the big impact of the bare second-order self energy is again removed if one considers the fully self consistent second-order scheme shown in Fig. 5.7b. An analytical argument shows that the corresponding spectral function of the renormalized propagator exhibits a singularity at zero frequency, i.e., $\rho(\omega) \propto |\omega|^{-\frac{1}{2}}$ for small ω, while numerical evaluation indicates a strong drop at large frequencies. This in turn evidences a spectral width much too small to generate a paramagnetic phase. Hence there is no simple way to control the fermion spectrum such that the approximation is well balanced between ordering and disordering effects. In order to cope with these difficulties we apply the FRG method in the following chapters since it sums up diagrammatic contributions in different interaction channels (either favoring order or disorder tendencies) in a systematic and self consistent way.

5 Finite Pseudo-Fermion Lifetime

6 The Functional Renormalization Group: Implementation for Spin Systems

Studying interacting fermion systems forms the core of todays condensed matter physics. Important representatives of such systems are the Hubbard or the t-J model which are widely believed to capture the relevant physics of high-T_c superconductivity. Also the fields of quantum dots, Kondo physics and Luttinger liquids fall into the class of problems described by interacting fermions. The general setup of these models is always given by a Hamiltonian H consisting of a kinetic term H_0 which is quadratic in the fermions and an interaction term H_{int} quartic in the fermions, $H = H_0 + H_{\text{int}}$. The progress in theoretical physics strongly depends on how accurately and reliably such systems can be calculated either analytically or numerically. Of special interest is the regime where H_{int} is not assumed small compared to H_0 giving rise to strong coupling phenomena. However, it is of course this regime which is not directly accessible using straightforward perturbation theory in H_{int}. The functional renormalization group (FRG) method [54, 80, 106, 142] is designed to handle interacting fermion systems beyond a simple perturbative expansion and helps to account for the interplay of different types of correlations controlling the physics of these models. The central concept of the FRG in general is a mapping of the initial Hamiltonian to a succession of Hamiltonians with gradually removed low energy degrees of freedom. This is realized by the introduction of a frequency cutoff Λ in the fermion propagator suppressing all modes with energies smaller than Λ. The initial Hamiltonian is reached in the limit $\Lambda \to 0$. Formulated in terms of an infinite hierarchy of differential equations for the m-particle vertex functions the FRG successively includes continuous energy modes during the flow of decreasing Λ and captures the related physics at that scale, until at the end of the flow ($\Lambda = 0$) the original model with its full dynamics is recovered. In principle the flow equations for the vertex functions generated by the FRG are exact but their numerical evaluation requires some truncation procedure which reduces the infinite hierarchy of equations to a finite set. However, even in truncated forms FRG includes systematic infinite order resummations of diagrams in different interaction channels as well as vertex corrections between them. In this way FRG is more powerful than straightforward renormalized perturbation theory since divergences that might occur in single channels are regularized. As a consequence FRG is in principle not restricted to small interaction strengths. So far, the range of applications comprises systems such as the two-dimensional Hubbard model [53, 58, 80, 95], Luttinger liquids with impurities [5, 6, 80], electron transport through correlated quantum dots [65],

6 The Functional Renormalization Group: Implementation for Spin Systems

the single impurity Anderson model [54, 66] and the Kondo dot model [108]. In the remaining part of this thesis we like to demonstrate that FRG is not only applicable to fermionic models but also to Heisenberg spin-systems. The essential ingredient for their implementation is again the pseudo-fermion representation introduced in Chapter 3.

This chapter is organized as follows: Section 6.1 presents the basic FRG formalism and sketches the derivation of the flow equations in the general form. The special implementation of the FRG for Heisenberg systems using pseudo fermions is then given in Section 6.2. After a brief discussion of a static FRG scheme in Section 6.3, the non-trivial problem of the truncation procedure of flow equations is discussed in Sections 6.4 and 6.5 where two different schemes are presented. Finally, Section 6.6 introduces a method that allows to characterize paramagnetic phases with respect to possible valence-bond dimerizations. In Sections 6.3, 6.4, 6.5 and 6.6 the approximation schemes under consideration are also applied to the J_1-J_2 model [98] in order to figure out their applicability.

6.1 General FRG Formalism

In this section we present the FRG formalism in its general form. The derivation of the flow equations is most conveniently formulated in the framework of generating functionals using the Feynman path-integral formalism. For a more detailed description we refer the reader to Refs. [54, 80, 106, 142]. The starting point is the path integral representation for the partition function of an interacting fermionic many-particle system

$$Z = \frac{1}{Z_0} \int D[\bar{\psi}, \psi] \exp[(\bar{\psi}, [G^0]^{-1}\psi) - H_{\text{int}}(\bar{\psi}, \psi)] . \quad (6.1)$$

Here $\bar{\psi}$ and ψ denote Grassmann fields associated with fermionic creation and annihilation operators. The corresponding Hamiltonian (expressed in terms of Grassmann fields) consists of a kinetic part

$$H_0(\bar{\psi}, \psi) = \sum_{k',k} \xi_{k',k} \bar{\psi}_{k'} \psi_k \quad (6.2)$$

which defines the free propagator G_0, and an interaction part

$$H_{\text{int}}(\bar{\psi}, \psi) = \sum_{k'_1, k'_2, k_1, k_2} V_{k'_1, k'_2, k_1, k_2} \bar{\psi}_{k'_1} \bar{\psi}_{k'_2} \psi_{k_2} \psi_{k_1} . \quad (6.3)$$

Correspondingly, the non-interacting partition function is defined as

$$Z_0 = \int D[\bar{\psi}, \psi] \exp[(\bar{\psi}, [G^0]^{-1}\psi)] . \quad (6.4)$$

All variables $k_i^{(\prime)}$ are understood as combined indices including the Matsubara frequencies and the quantum numbers in which the problem is formulated. In addition, we use the short hand notation

$$(\bar{\psi}, [G^0]^{-1}\psi) = \sum_{k',k} \bar{\psi}_{k'} [G^0]^{-1}_{k',k} \psi_k . \quad (6.5)$$

Note that due to energy conservation G^0 is diagonal in the Matsubara frequencies but not necessarily diagonal in the quantum numbers as they are kept unspecified here. We obtain the generating functional for the dressed, disconnected m-particle Green's functions by adding source terms in the exponent,

$$W(\bar{\eta},\eta) = \frac{1}{Z_0}\int D[\bar{\psi},\psi]\exp[(\bar{\psi},[G^0]^{-1}\psi) - H_{\text{int}}(\bar{\psi},\psi) - (\bar{\psi},\eta) - (\bar{\eta},\psi)]. \quad (6.6)$$

Taking the logarithm we gain the generating functional for connected m-particle Green's functions,

$$W^c(\bar{\eta},\eta) = \ln[W(\bar{\eta},\eta)]. \quad (6.7)$$

The connected Green's functions themselves are generated by applying functional derivatives,

$$G^c_m(k'_1,\ldots,k'_m;k_1,\ldots,k_m) = \frac{\delta^m}{\delta\bar{\eta}_{k'_1}\cdots\delta\bar{\eta}_{k'_m}}\frac{\delta^m}{\delta\eta_{k_m}\cdots\delta\eta_{k_1}}W^c(\bar{\eta},\eta)\bigg|_{\bar{\eta}=\eta=0}. \quad (6.8)$$

Introducing new Grassmann fields

$$\phi = -\frac{\delta W^c(\bar{\eta},\eta)}{\delta\bar{\eta}} \quad \text{and} \quad \bar{\phi} = \frac{\delta W^c(\bar{\eta},\eta)}{\delta\eta} \quad (6.9)$$

and performing a Legendre transformation, we obtain the generating functional for the one-particle irreducible (1PI) vertex functions

$$\Gamma(\bar{\phi},\phi) = -W^c(\bar{\eta},\eta) - (\bar{\phi},\eta) - (\bar{\eta},\phi) + (\bar{\phi},[G^0]^{-1}\phi). \quad (6.10)$$

For the vertex functions γ_m an expression similar to Eq. (6.8) holds,

$$\gamma_m(k'_1,\ldots,k'_m;k_1,\ldots,k_m) = \frac{\delta^m}{\delta\bar{\phi}_{k'_1}\cdots\delta\bar{\phi}_{k'_m}}\frac{\delta^m}{\delta\phi_{k_m}\cdots\delta\phi_{k_1}}\Gamma(\bar{\phi},\phi)\bigg|_{\bar{\phi}=\phi=0}. \quad (6.11)$$

These are the basic objects of the one-particle irreducible version of FRG. For the derivation of the flow equations we need a relation between the second derivatives of W^c and Γ. Simple algebra yields

$$\begin{pmatrix} \frac{\delta^2 W^c}{\delta\bar{\eta}\delta\eta} & -\frac{\delta^2 W^c}{\delta\bar{\eta}\delta\bar{\eta}} \\ -\frac{\delta^2 W^c}{\delta\eta\delta\eta} & \frac{\delta^2 W^c}{\delta\eta\delta\bar{\eta}} \end{pmatrix} = \begin{pmatrix} \frac{\delta^2\Gamma}{\delta\bar{\phi}\delta\phi}+[G^0]^{-1} & \frac{\delta^2\Gamma}{\delta\bar{\phi}\delta\bar{\phi}} \\ \frac{\delta^2\Gamma}{\delta\phi\delta\phi} & \frac{\delta^2\Gamma}{\delta\phi\delta\bar{\phi}}-[[G^0]^{-1}]^{\text{T}} \end{pmatrix}^{-1}. \quad (6.12)$$

From this identity the physical meaning of γ_1 can be extracted. Putting all source fields to zero and assuming that the system is not in a symmetry-broken phase (i.e., terms of the form $\frac{\delta^2 W^c}{\delta\eta\delta\eta}$ vanish) the upper left element of Eq. (6.12) reads

$$G \equiv G_1 = \frac{\delta^2 W^c}{\delta\bar{\eta}\delta\eta}\bigg|_{\bar{\eta}=\eta=0} = \left[\frac{\delta^2\Gamma}{\delta\bar{\phi}\delta\phi}\bigg|_{\bar{\phi}=\phi=0}+[G^0]^{-1}\right]^{-1} = [\gamma_1+[G^0]^{-1}]^{-1}. \quad (6.13)$$

A comparison with Dyson's equation $G = [[G^0]^{-1} - \Sigma]^{-1}$ shows that γ_1 is related to the self energy Σ via

$$\gamma_1 = -\Sigma. \tag{6.14}$$

As already pointed out, the FRG procedure amounts to introducing a (Matsubara-) frequency cutoff in the bare Green's function,

$$G^{0\Lambda}(\omega) = \Theta(|\omega| - \Lambda)G^0(\omega), \tag{6.15}$$

which suppresses all modes below the energy scale Λ. In principle, the cutoff procedure may also be performed in momentum space, however, as will be explained in the next section, for an application to spin systems a frequency cutoff is more suitable. There is also some freedom concerning the explicit form of the cutoff function. The only restriction is that $G^{0\Lambda} \equiv 0$ in the limit $\Lambda \to \infty$ and $G^{0\Lambda} = G^0$ at $\Lambda = 0$. In Eq. 6.15 and also in the following sections we choose a sharp cutoff function which has the advantage that internal frequency integrations are canceled in the flow equations. There are, however, situations where a broadened cutoff might be a better choice, e.g. at finite temperatures, see Chapter 9.

The FRG formulates differential equations for the 1PI vertex functions under the flow of Λ. In the following we will derive these equations. The starting point is the generating functional for the connected Green's functions W^c which becomes Λ-dependent upon replacing G^0 by $G^{0\Lambda}$. We calculate its Λ-derivative,

$$\frac{d}{d\Lambda}W^{c\Lambda} = -\text{Tr}[Q^\Lambda G^{0\Lambda}] + \text{Tr}\left[Q^\Lambda \frac{\delta^2 W^{c\Lambda}}{\delta\bar\eta\delta\eta}\right] - \left(\frac{\delta W^{c\Lambda}}{\delta\eta}, Q^\Lambda \frac{\delta W^{c\Lambda}}{\delta\bar\eta}\right), \tag{6.16}$$

where we defined

$$Q^\Lambda = \frac{d}{d\Lambda}[G^{0\Lambda}]^{-1}. \tag{6.17}$$

We are actually interested in the Λ-derivative of the generating functional for the 1PI vertex function Γ^Λ. Using Eqs. (6.10) and (6.16) and taking notice of the fact that $\bar\phi$ and ϕ are the fundamental variables of Γ^Λ we arrive at

$$\frac{d}{d\Lambda}\Gamma^\Lambda = \text{Tr}[Q^\Lambda G^{0\Lambda}] - \text{Tr}\left[Q^\Lambda \frac{\delta^2 W^c}{\delta\bar\eta^\Lambda \delta\eta^\Lambda}\right]. \tag{6.18}$$

We emphasize that due to the change of variables in the Legendre transformation the old variables $\bar\eta$ and η acquire a Λ-dependence. Employing Eq. (6.12) we are now able to eliminate W^c in the above differential equation,

$$\frac{d}{d\Lambda}\Gamma^\Lambda = \text{Tr}[Q^\Lambda G^{0\Lambda}] - \text{Tr}[G^\Lambda Q^\Lambda R_{11}]. \tag{6.19}$$

Here R_{11} denotes the upper left matrix element of

$$R = \left[\mathbb{1} - \begin{pmatrix} -G^\Lambda & 0 \\ 0 & [G^\Lambda]^T \end{pmatrix} \begin{pmatrix} U & \frac{\delta^2 \Gamma^\Lambda}{\delta\phi\delta\phi} \\ \frac{\delta^2 \Gamma^\Lambda}{\delta\bar\phi\delta\bar\phi} & -U^T \end{pmatrix}\right]^{-1} \tag{6.20}$$

with
$$U = \frac{\delta^2 \Gamma^\Lambda}{\delta\bar{\phi}\delta\phi} - \gamma_1^\Lambda. \tag{6.21}$$

The final step in our derivation is the expansion of Γ^Λ in powers of $\bar{\phi}$ and ϕ,

$$\Gamma^\Lambda(\bar{\phi},\phi) = \sum_{m=0}^{\infty} \frac{(-1)^m}{(m!)^2} \sum_{k_1',\ldots,k_m'} \sum_{k_1,\ldots,k_m} \gamma_m^\Lambda(k_1',\ldots,k_m';k_1,\ldots,k_m)\bar{\phi}_{k_1'}\cdots\bar{\phi}_{k_m'}\phi_{k_m}\cdots\phi_{k_1}. \tag{6.22}$$

Expanding R from Eq. (6.20) in a geometric series and comparing the coefficients of the fields $\bar{\phi}$ and ϕ on both sides of Eq. (6.19) we finally obtain the flow equations for γ_i^Λ. We write down the equations for γ_1^Λ and γ_2^Λ,

$$\frac{d}{d\Lambda}\gamma_1^\Lambda(k_1',k_1) = \frac{1}{\beta}\sum_{k_2',k_2}\gamma_2^\Lambda(k_1',k_2';k_1,k_2)S^\Lambda(k_2,k_2'), \tag{6.23a}$$

$$\frac{d}{d\Lambda}\gamma_2^\Lambda(k_1',k_2';k_1,k_2) = \frac{1}{\beta}\sum_{k_3',k_3}\gamma_3^\Lambda(k_1',k_2',k_3';k_1,k_2,k_3)S^\Lambda(k_3,k_3')$$
$$+ \frac{1}{\beta}\sum_{k_3',k_3}\sum_{k_4',k_4}[\gamma_2^\Lambda(k_1',k_2';k_3,k_4)\gamma_2^\Lambda(k_3',k_4';k_1,k_2)$$
$$- \gamma_2^\Lambda(k_1',k_4';k_1,k_3)\gamma_2^\Lambda(k_3',k_2';k_4,k_2) - (k_3' \leftrightarrow k_4', k_3 \leftrightarrow k_4)$$
$$+ \gamma_2^\Lambda(k_2',k_4';k_1,k_3)\gamma_2^\Lambda(k_3',k_1';k_4,k_2) + (k_3' \leftrightarrow k_4', k_3 \leftrightarrow k_4)]$$
$$\times G^\Lambda(k_3,k_3')S^\Lambda(k_4,k_4'). \tag{6.23b}$$

In this representation we explicitly wrote the traces as sums. The prefactors $\frac{1}{\beta} = T$ are associated with Matsubara sums. Due to energy conservation only one Matsubara sum remains on the right hand side. Furthermore we have introduced the so-called single-scale propagator

$$S^\Lambda = G^\Lambda Q^\Lambda G^\Lambda. \tag{6.24}$$

Eqs. (6.23a) and (6.23b) show that the flow of γ_1^Λ is determined by γ_2^Λ and γ_1^Λ, where the dependence on the latter is included in the Green's function. In turn, the flow of γ_2^Λ depends on γ_3^Λ, γ_2^Λ and γ_1^Λ. This scheme applies to arbitrary γ_i^Λ: The flow equation for γ_i^Λ contains all γ_j^Λ with $j \leq i+1$ such that the hierarchy of equations never closes.

In Fig. 6.1 these equations are depicted in diagrammatic form. The right hand side of the flow equation for the two-particle vertex γ_2^Λ consists of six terms (which appear in the same order as in Eq. (6.23b)): The first one is the contribution of the three particle vertex. All approximation schemes used in the following apply to this term. The second is the so-called particle-particle graph while the remaining four terms are referred to as particle-hole graphs. Concerning these four diagrams one again distinguishes between the direct particle-hole terms (third and fourth term) and the crossed particle-hole terms (fifth and sixth term).

We still have to specify the initial conditions which can be deduced within a simple argument. In the limit $\Lambda \to \infty$ the Green's function $G^{0\Lambda}$ vanishes and consequently

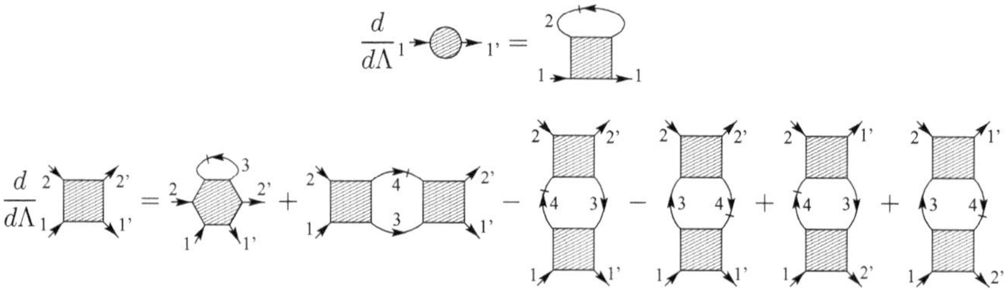

6 The Functional Renormalization Group: Implementation for Spin Systems

Figure 6.1: FRG equations for γ_1^Λ (first line) and γ_2^Λ (second line). The line with an arrow is the full Green's function $G^\Lambda(\omega)$ and the line with an arrow and a slash denotes the single-scale propagator $S^\Lambda(\omega)$. All variables are presented as numbers.

particle propagation is turned off. Hence, in diagrammatic expressions only the bare vertices remain such that the initial conditions are given by

$$\gamma_1^{\Lambda \to \infty}(k', k) = -\Sigma(k', k) = -\xi_{k',k}, \tag{6.25a}$$

$$\gamma_2^{\Lambda \to \infty}(k'_1, k'_2; k_1, k_2) = V_{k'_1, k'_2, k_1, k_2}, \tag{6.25b}$$

$$\gamma_m^{\Lambda \to \infty}(k'_1, \ldots, k'_m; k_1, \ldots, k_m) = 0 \text{ for } m \geq 3. \tag{6.25c}$$

The single-particle potential $\xi_{k',k}$ and the two-particle interaction $V_{k'_1,k'_2,k_1,k_2}$ have been introduced in Eqs. (6.2) and (6.3), respectively. For the third condition (6.25c) we assumed that the Hamiltonian does not contain any higher interactions than two-particle scattering terms. At the end of the flow at $\Lambda = 0$ when the theory is cutoff free, the exact vertex functions are obtained.

So far we have presented the most general form of the FRG formalism. All steps of implementation specific to spin systems are considered in the next section.

6.2 FRG and its Implementation for Heisenberg Systems

An FRG scheme for spin systems using pseudo fermions necessitates some basic methodological modifications compared to the standard electronic many-particle systems mentioned at the beginning of this chapter. Unlike conventional FRG approaches, our starting point is not given by the bare excitations of the system but is rather based on auxiliary degrees of freedom. In the spin representation (3.1) the pseudo fermions carry quantum numbers in the form of a site index i and a spin index α. Most naturally, the FRG is formulated in terms of these variables because in real-space representation all propagators are local entities. We like to stress that this is actually the first time FRG is applied in real space.

In principle, in order to describe RG flows into long-range ordered phases it is necessary to include small symmetry breaking fields in the Hamiltonian. In this chapter we constrain ourselves to non-magnetic phases and defer consideration of the flow of

6 The Functional Renormalization Group: Implementation for Spin Systems

the magnetic order parameter to Chapter 8. This has the advantage that for the spin structure of the vertex functions only rotation invariant forms have to be taken into account. Note that although within this scheme, the magnetic phases are not directly accessible, magnetic instabilities may be detected as a breakdown of the flow.

From these considerations one may write Eqs. (6.23a) and (6.23b) in a more convenient form,

$$\frac{d}{d\Lambda}\Sigma^\Lambda(\omega_1) = -\frac{1}{2\pi}\sum_2 \Gamma^\Lambda(1,2;1,2)S^\Lambda(\omega_2), \quad (6.26a)$$

$$\frac{d}{d\Lambda}\Gamma^\Lambda(1',2';1,2) = \frac{1}{2\pi}\sum_{3,4}[\Gamma^\Lambda(1',2';3,4)\Gamma^\Lambda(3,4;1,2)$$
$$-\Gamma^\Lambda(1',4;1,3)\Gamma^\Lambda(3,2';4,2) - (3\leftrightarrow 4)$$
$$+\Gamma^\Lambda(2',4;1,3)\Gamma^\Lambda(3,1';4,2) + (3\leftrightarrow 4)]G^\Lambda(\omega_3)S^\Lambda(\omega_4). \quad (6.26b)$$

To shorten the notation with respect to the indices, here and in the following we write the two-particle vertex as Γ^Λ instead of γ_2^Λ (this should not be confused with the generating functional $\Gamma^\Lambda(\bar\phi,\phi)$). Also the first flow equation is formulated for the self energy Σ^Λ rather than $\gamma_1^\Lambda = -\Sigma^\Lambda$. The numbers are shorthand notations for the frequency, the site index and the spin index, that is, $1 = \{\omega_1, i_1, \alpha_1\}$. In the following we omit the imaginary unit i in frequency arguments of propagators and vertex functions; they are always understood as Matsubara frequencies. Furthermore, we use that Σ^Λ, G^Λ and S^Λ are local in real space and proportional to the unity matrix in spin space. If we assume that all lattice sites are equivalent (as it is the case for all models considered in this thesis) the one-particle quantities Σ^Λ, G^Λ and S^Λ do not carry a site index. That is, we may write

$$\Sigma^\Lambda(1,1') = \Sigma^\Lambda(\omega_1)\delta(\omega_1-\omega_{1'})\delta_{i_1 i_{1'}}\delta_{\alpha_1,\alpha_{1'}}, \quad (6.27)$$

and similarly for G^Λ and S^Λ. Using these relations some summations on the right hand side of Eq. (6.26b) cancel. In Eqs. (6.26a) and (6.26b) the limit $T \to 0$ has already been performed, converting the discrete Matsubara sums into integrals and the factors $\frac{1}{\beta}$ into $\frac{d\omega}{2\pi}$. That is, Σ_1 stands for an integral over ω_1 and sums over i_1 and α_1.

It will turn out that the truncation procedure of the flow equations is a non-trivial problem in our case. In order to obtain a closed and finite set of equations we either neglect the three-particle vertex γ_3^Λ completely (see the conventional FRG scheme in Section 6.4) or keep particular contributions of it (see the Katanin scheme in Section 6.5). However, in the latter scheme it is possible to absorb the contribution of γ_3^Λ into a redefined single-scale propagator S^Λ. In either case, one may write the second flow equation in the form of Eq. (6.26b) where γ_3^Λ does not explicitly appear.

It is an unusual property of our FRG scheme that a kinetic term is absent in the Hamiltonian which implies that we operate in the strong-coupling limit. Nevertheless the derivation of the flow equations presented in the previous section is also valid in our case and one can apply the FRG scheme in the usual way. The bare, scale dependent

Green's function is given by
$$G^{0\Lambda}(\omega) = \frac{\Theta(|\omega| - \Lambda)}{i\omega + \mu}. \tag{6.28}$$

If not stated otherwise we employ the average projection scheme and set $\mu = 0$. However, it is not too difficult to implement the exact projection scheme [91], which increases the numerical effort by roughly a factor of 16. Although the self energy $\Sigma^\Lambda(\omega)$ vanishes in the initial conditions, it becomes finite during the flow. Hence, the full propagator $G^\Lambda(\omega)$ reads
$$G^\Lambda(\omega) = \frac{\Theta(|\omega| - \Lambda)}{i\omega - \Sigma^\Lambda(\omega)}. \tag{6.29}$$

Using a sharp cutoff function a technical difficulty arises regarding the single-scale propagator,
$$S^\Lambda(\omega) = \left(G^\Lambda(\omega)\right)^2 \frac{d}{d\Lambda} \left(G^{0\Lambda}(\omega)\right)^{-1}. \tag{6.30}$$

This expression contains the step function $\Theta(|\omega| - \Lambda)$ as well as the δ-peak $\delta(|\omega| - \Lambda)$ (which is generated by the Λ-derivative). Since both functions are non-analytic at the same points $\omega = \pm\Lambda$ such an expression seems to be ambiguous at a first glance. However, this ambiguity can be resolved using broadened functions δ_ϵ and Θ_ϵ with the broadening parameter ϵ and subsequently considering the limit $\epsilon \to 0$. In this way an identity introduced by Morris [82] may be proved,
$$\delta_\epsilon(x - \Lambda) f(\Theta_\epsilon(x - \Lambda)) \to \delta(x - \Lambda) \int_0^1 f(t) dt, \tag{6.31}$$

where f is an arbitrary continuous function. Using this identity the right hand side of Eq. (6.30) may be evaluated, yielding
$$S^\Lambda(\omega) = \frac{\delta(|\omega| - \Lambda)}{i\omega - \Sigma^\Lambda(\omega)}. \tag{6.32}$$

Now the advantage of a sharp cutoff becomes obvious: The δ-function removes the frequency integrations on the right hand sides of Eqs. (6.26a) and (6.26b) such that (at least for the conventional truncation scheme) no integrations remain in the first two flow equations.

Initial Conditions

Next we specify the initial conditions at $\Lambda \to \infty$. We have already stated that the self energy vanishes in this limit,
$$\Sigma^{\Lambda \to \infty} \equiv 0. \tag{6.33}$$

For the initial conditions of the two-particle vertex we have to take into account that due to the anti-commutativity of the fermions, the vertex $\Gamma^\Lambda(1', 2'; 1, 2)$ changes sign under an exchange of the variables $1 \leftrightarrow 2$ or $1' \leftrightarrow 2'$. Thus, the bare interaction in antisymmetrized form is given by
$$\Gamma^{\Lambda \to \infty}(1', 2'; 1, 2) = J_{i_1 i_2} \frac{1}{2}\sigma^\mu_{\alpha_{1'} \alpha_1} \frac{1}{2}\sigma^\mu_{\alpha_{2'} \alpha_2} \delta_{i_{1'} i_1} \delta_{i_{2'} i_2} - J_{i_1 i_2} \frac{1}{2}\sigma^\mu_{\alpha_{1'} \alpha_2} \frac{1}{2}\sigma^\mu_{\alpha_{2'} \alpha_1} \delta_{i_{1'} i_2} \delta_{i_{2'} i_1}. \tag{6.34}$$

Here the factors $\frac{1}{2}\sigma^\mu_{\alpha\beta}$ originate from the spin representation (3.1) and the Kronecker δ ensure that there is no fermion hopping on the lattice.

Parametrization of the Vertices

In order to proceed, the vertex functions Σ^Λ and Γ^Λ have to be parametrized. Firstly we know that the self energy is a local quantity proportional to the unity matrix in spin space. As a consequence of particle-hole symmetry and time-reversal symmetry, the self energy is an odd function with vanishing real part along the imaginary Matsubara axis, as already pointed out after Eq. (5.4). In analogy to Eq. (5.4) we write

$$\Sigma^\Lambda(\omega) = -i\gamma^\Lambda(\omega). \tag{6.35}$$

Since the rotational invariance in spin space of the initial conditions is conserved during the flow, the two-particle vertex at finite Λ is parametrized by spin-interaction terms $\propto \sigma^\mu_{\alpha\beta}\sigma^\mu_{\gamma\delta}$ and density-interaction terms $\propto \delta_{\alpha\beta}\delta_{\gamma\delta}$. These are the only matrix combinations in spin space which preserve rotational invariance. At first, it is not obvious why density-interaction terms should be relevant in pure spin models. Nevertheless, these terms play an important role in our FRG scheme. Since the propagators are local, the site index of an ingoing leg has to be identical to the site index of the corresponding outgoing leg, which results in a total dependence on only two sites, i.e., i_1 and i_2. To be more precise, translation invariance of the underlying lattice further reduces the site dependence only to the separation $|\mathbf{R}_{i_1} - \mathbf{R}_{i_2}|$ (we note that in the case of non-Bravais like lattices as discussed in Sections 7.2, 7.4, 7.5 and 7.6 a closer investigation of the site-dependence of the two-particle vertex is necessary, see the discussion in Section 7.2). Taking into account the antisymmetry in all variables the two-particle vertex can now be represented as

$$\begin{aligned}\Gamma^\Lambda(1',2';1,2) = \{ &[\Gamma^\Lambda_{s\,i_1 i_2}(\omega_{1'},\omega_{2'};\omega_1,\omega_2)\sigma^\mu_{\alpha_{1'}\alpha_1}\sigma^\mu_{\alpha_{2'}\alpha_2} + \Gamma^\Lambda_{d\,i_1 i_2}(\omega_{1'},\omega_{2'};\omega_1,\omega_2)\delta_{\alpha_{1'}\alpha_1}\delta_{\alpha_{2'}\alpha_2}]\\ &\times \delta_{i_{1'}i_1}\delta_{i_{2'}i_2}\\ &- [\Gamma^\Lambda_{s\,i_1 i_2}(\omega_{1'},\omega_{2'};\omega_2,\omega_1)\sigma^\mu_{\alpha_{1'}\alpha_2}\sigma^\mu_{\alpha_{2'}\alpha_1} + \Gamma^\Lambda_{d\,i_1 i_2}(\omega_{1'},\omega_{2'};\omega_2,\omega_1)\delta_{\alpha_{1'}\alpha_2}\delta_{\alpha_{2'}\alpha_1}]\\ &\times \delta_{i_{1'}i_2}\delta_{i_{2'}i_1}\}\delta(\omega_1+\omega_2-\omega_{1'}-\omega_{2'}). \end{aligned} \tag{6.36}$$

The indices s/d correspond to spin and density interactions, respectively. The δ-function $\delta(\omega_1+\omega_2-\omega_{1'}-\omega_{2'})$ ensures the fulfillment of energy conservation.

Explicit Flow Equations

We are now prepared to calculate the flow equations for γ^Λ, Γ^Λ_s and Γ^Λ_d. We start with the equation for the self energy γ^Λ. Inserting Eqs. (6.32), (6.35) and (6.36) and into Eq. (6.26a) and performing the ω_2-integration and the spin sums we arrive at

$$\frac{d}{d\Lambda}\gamma^\Lambda(\omega_1) = \frac{1}{2\pi}\sum_{\omega_2=\pm\Lambda}\left[-2\sum_{i_2}\Gamma^\Lambda_{d\,i_1 i_2}(\omega_1,\omega_2;\omega_1,\omega_2) + 3\Gamma^\Lambda_{s\,i_1 i_1}(\omega_1,\omega_2;\omega_2,\omega_1)\right.$$
$$\left.+ \Gamma^\Lambda_{d\,i_1 i_1}(\omega_1,\omega_2;\omega_2,\omega_1)\right]\frac{1}{\omega_2+\gamma^\Lambda(\omega_2)}. \tag{6.37}$$

6 The Functional Renormalization Group: Implementation for Spin Systems

Later it will turn out that regarding the second flow equation some symmetries in the frequencies may be exploited if one transforms the frequency arguments of Γ_s^Λ and Γ_d^Λ as follows,

$$\Gamma_{s/d\, i_1 i_2}^\Lambda(\omega_1, \omega_2; \omega_3, \omega_4) \to \Gamma_{s/d\, i_1 i_2}^\Lambda(\omega_1 + \omega_2, \omega_1 - \omega_3, \omega_1 - \omega_4). \tag{6.38}$$

Note that due to energy conservation three frequency variables are sufficient. Especially for the external frequencies $\omega_1, \omega_{1'}, \omega_2, \omega_{2'}$ we introduce the invariant variables s, t, u defined by

$$s = \omega_{1'} + \omega_{2'}, \quad t = \omega_{1'} - \omega_1, \quad u = \omega_{1'} - \omega_2. \tag{6.39}$$

Rewriting Eq. (6.37) in terms of these new variables and using $\gamma^\Lambda(\omega) = -\gamma^\Lambda(-\omega)$ we obtain our final version of the first flow equation,

$$\frac{d}{d\Lambda}\gamma^\Lambda(\omega) = \frac{1}{2\pi}\Bigg[-2\sum_j \left(\Gamma_{d\,ij}^\Lambda(\omega + \Lambda, 0, \omega - \Lambda) - \Gamma_{d\,ij}^\Lambda(\omega - \Lambda, 0, \omega + \Lambda)\right)$$
$$+ 3\left(\Gamma_{s\,ii}^\Lambda(\omega + \Lambda, \omega - \Lambda, 0) - \Gamma_{s\,ii}^\Lambda(\omega - \Lambda, \omega + \Lambda, 0)\right)$$
$$+ \Gamma_{d\,ii}^\Lambda(\omega + \Lambda, \omega - \Lambda, 0) - \Gamma_{d\,ii}^\Lambda(\omega - \Lambda, \omega + \Lambda, 0)\Bigg]\frac{1}{\Lambda + \gamma^\Lambda(\Lambda)}. \tag{6.40}$$

We emphasize that this equation holds for an arbitrary reference site i.

Next we sketch the derivation of the differential equations for Γ_s^Λ and Γ_d^Λ. For the sake of brevity we use the notation

$$\Gamma^\Lambda(1', 2'; 1, 2) = \Gamma_{=\,i_1 i_2}^\Lambda(1', 2'; 1, 2)\delta_{i_{1'} i_1}\delta_{i_{2'} i_2} - \Gamma_{\times\,i_1 i_2}^\Lambda(1', 2'; 1, 2)\delta_{i_{1'} i_2}\delta_{i_{2'} i_1}, \tag{6.41}$$

where in correspondence to Eq. (6.36) $\Gamma_=^\Lambda$ and $\Gamma_\times^\Lambda$ are defined by

$$\Gamma_{=\,i_1 i_2}^\Lambda(1', 2'; 1, 2) = \left[\Gamma_{s\,i_1 i_2}^\Lambda(\omega_{1'}, \omega_{2'}; \omega_1, \omega_2)\sigma_{\alpha_{1'}\alpha_1}^\mu \sigma_{\alpha_{2'}\alpha_2}^\mu + \Gamma_{d\,i_1 i_2}^\Lambda(\omega_{1'}, \omega_{2'}; \omega_1, \omega_2)\delta_{\alpha_{1'}\alpha_1}\delta_{\alpha_{2'}\alpha_2}\right]$$
$$\times \delta(\omega_1 + \omega_2 - \omega_{1'} - \omega_{2'}), \tag{6.42a}$$

$$\Gamma_{\times\,i_1 i_2}^\Lambda(1', 2'; 1, 2) = \left[\Gamma_{s\,i_1 i_2}^\Lambda(\omega_{1'}, \omega_{2'}; \omega_2, \omega_1)\sigma_{\alpha_{1'}\alpha_2}^\mu \sigma_{\alpha_{2'}\alpha_1}^\mu + \Gamma_{d\,i_1 i_2}^\Lambda(\omega_{1'}, \omega_{2'}; \omega_2, \omega_1)\delta_{\alpha_{1'}\alpha_2}\delta_{\alpha_{2'}\alpha_1}\right]$$
$$\times \delta(\omega_1 + \omega_2 - \omega_{1'} - \omega_{2'}). \tag{6.42b}$$

These quantities indicate the two possibilities of connecting the external fermion lines of the two-particle vertex. In diagrammatic representation this may be displayed as

$$\Gamma_{=\,i_1 i_2}^\Lambda(1', 2'; 1, 2) = \begin{array}{c}\text{2}\!\!\!\longrightarrow\!\!\!\text{2'}\\ \text{1}\!\!\!\longrightarrow\!\!\!\text{1'}\end{array}, \quad \Gamma_{\times\,i_1 i_2}^\Lambda(1', 2'; 1, 2) = \begin{array}{c}\text{2}\!\!\!\longrightarrow\!\!\!\text{2'}\\ \text{1}\!\!\!\longrightarrow\!\!\!\text{1'}\end{array}, \tag{6.43}$$

where the thick fermion lines are again understood as lines of constant lattice site. Note that the arguments of $\Gamma_=^\Lambda$ and $\Gamma_\times^\Lambda$ indicated by numbers are composite indices comprising only frequency and spin, i.e., $1 = \{\omega_1, \alpha_1\}$. There is a simple relation between $\Gamma_=^\Lambda$ and $\Gamma_\times^\Lambda$,

$$\Gamma_{=\,i_1 i_2}^\Lambda(1', 2'; 1, 2) = \Gamma_{\times\,i_2 i_1}^\Lambda(1', 2'; 2, 1). \tag{6.44}$$

6 The Functional Renormalization Group: Implementation for Spin Systems

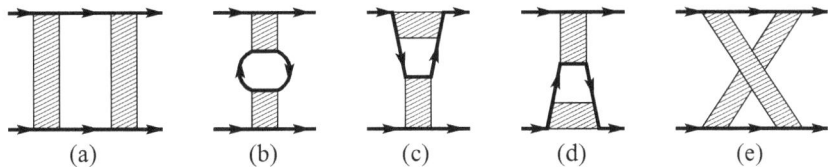

Figure 6.2: The contributions to the right hand side of Eq. (6.45). Thick fermion propagators represent lines of constant lattice site. Diagram (a) is the particle-particle term, (b), (c) and (d) are direct particle-hole terms and (e) is the crossed particle-hole term. The order of terms is the same as in Eq. (6.45).

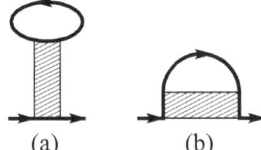

Figure 6.3: The contributions to the self energy, see Eq. (6.40). Diagram (a) represents the Hartree term, (b) the Fock term.

Rewriting Eq. (6.26b) in terms of $\Gamma^\Lambda_=$ and $\Gamma^\Lambda_\times$ the site summations can be easily performed. Comparing the coefficients of $\delta_{i_1'i_1}\delta_{i_2'i_2}$ on both sides of this equation and using Eq. (6.44) a differential equation for $\Gamma^\Lambda_=$ is obtained,

$$\frac{d}{d\Lambda}\Gamma^\Lambda_{=i_1i_2}(1',2';1,2) = \frac{1}{2\pi}\int_{-\infty}^{\infty}d\omega_4\int_{-\infty}^{\infty}d\omega_3\sum_{\alpha_3\alpha_4}$$
$$\Big[\ \Gamma^\Lambda_{=i_1i_2}(1',2';3,4)\Gamma^\Lambda_{=i_1i_2}(3,4;1,2) + (3\leftrightarrow 4)$$
$$+\sum_j\Gamma^\Lambda_{=i_1j}(1',4;1,3)\Gamma^\Lambda_{=ji_2}(3,2';4,2) + (3\leftrightarrow 4)$$
$$-\Gamma^\Lambda_{=i_1i_2}(1',4;1,3)\Gamma^\Lambda_{=i_2i_2}(3,2';2,4) - (3\leftrightarrow 4)$$
$$-\Gamma^\Lambda_{=i_1i_1}(1',4;3,1)\Gamma^\Lambda_{=i_1i_2}(3,2';4,2) - (3\leftrightarrow 4)$$
$$+\Gamma^\Lambda_{=i_2i_1}(2',4;3,1)\Gamma^\Lambda_{=i_2i_1}(3,1';2,4) + (3\leftrightarrow 4)\Big]$$
$$\times G^\Lambda(\omega_3)S^\Lambda(\omega_4)\,. \qquad (6.45)$$

The five terms in Eq. (6.45) have a graphical representation shown in Fig. 6.2. The first term (a) is the particle-particle term generating ladder-like diagrams with fermion lines of the same direction. The three contributions (b), (c) and (d) are the so-called direct particle-hole terms while the last term (e) is the crossed particle-hole graph generating ladder-like diagrams with fermion lines of the opposite orientation. A special role plays the term (b) because it is the only contribution that exhibits a closed internal fermion loop. This bubble can be located on an arbitrary lattice site leading to the site summation in the second line of Eq. (6.45). Consequently, this is the only term for which a vertex with a given spatial distance $\mathbf{R}_{i_1} - \mathbf{R}_{i_2}$ on the left hand side of Eq. (6.45) couples to vertices with all possible distance vectors on the right hand side. For all other terms the vertex on the left hand side only couples to vertices with the same lattice indices (or in case of diagrams (c) and (d) also to the on-site vertex). It is the mutual coupling of all vertices generated by term (b) which makes this contribution essential for spin models because this property allows to describe spin fluctuations throughout the lattice possibly leading to collective, long-range ordering phenomena. On the other hand, the remaining terms (a), (c), (d) and (e) are not able to induce long-range order. Since the bubble structure of term (b) generates RPA-like diagrams, we refer to it as the RPA-term.

Similar pictures may be drawn for the right hand side of the first flow equation (6.40), see Fig. 6.3. Here the sum term in the first line of Eq. (6.40) can be identified with the Hartree term in Fig. 6.3a. The on-site vertices in the other lines correspond to the Fock graph (b).

The last steps of derivation towards flow equations for Γ_s^Λ and Γ_d^Λ amount to inserting Eq. (6.42a) into Eq. (6.45). The final version of the equations is presented in appendix B. In the following we study some characteristics of the flow equations (6.40), (B.1) and (B.2) and describe the numerical implementation. At this stage we leave the truncation scheme unspecified since the two procedures considered in this thesis only affect the definition of the single-scale propagator: Within the conventional truncation, see Section 6.4 where the three-particle vertex is completely discarded the standard definition (6.32) holds while within the Katanin scheme the single-scale propagator appearing in the second flow equation (6.45) acquires an extra term that contains the derivative of the self energy, $\frac{d}{d\Lambda}\gamma^\Lambda$, see Section 6.5. Here we also state the initial conditions for γ^Λ, Γ_s^Λ and Γ_d^Λ which are deduced comparing Eqs. (6.34) and (6.36),

$$\gamma^{\Lambda\to\infty}(\omega) = 0,$$
$$\Gamma_{s\,i_1 i_2}^{\Lambda\to\infty}(s,t,u) = \tfrac{1}{4}J_{i_1 i_2}, \quad \Gamma_{d\,i_1 i_2}^{\Lambda\to\infty}(s,t,u) = 0. \qquad (6.46)$$

Approximating the Spatial Dependence

In general, we are dealing with infinite lattices leading to an infinitely large set of equations for $\Gamma_{s\,i_1 i_2}^\Lambda$ and $\Gamma_{d\,i_1 i_2}^\Lambda$ with all possible indices i_1 and i_2. Since the numerical solution requires a finite set, the spatial dependence has to be approximated keeping two-particle vertices only up to a finite distance between i_1 and i_2 and discarding longer vertices. The term in Fig. 6.3a as well as the term in Fig. 6.2b contain site summations (also seen in Eqs. (6.40), (B.1) and (B.2)) which will also be affected by the real-space cutoff scheme. We discuss three different ways of implementing such a scheme and comment on their advantages and disadvantages:

(1) **Hard-wall boundary conditions:** In principle one might consider finite systems with hard-wall boundary conditions. However, this complicates the numerical effort enormously as compared to schemes where translation invariance is intact. The loss of translational invariance means that the self energy carries a site index and consequently Eq. (6.40) has to be calculated for each site separately. Additionally, in such a scheme the two-particle vertex would depend on two site indices i_1 and i_2 individually instead of just the distance vector between them, leading to

6 The Functional Renormalization Group: Implementation for Spin Systems

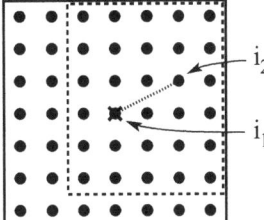

Figure 6.4: Array of 7×7 sites to illustrate different approximation schemes for the spatial dependence of $\Gamma^\Lambda_{\text{s/d}\, i_1 i_2}$. For details, see text.

equations for all combinations of pairs (i_1, i_2) in the finite system (lattice symmetries may, however, reduce the effort a bit). For these reasons we will not consider finite systems, but rather turn to schemes that preserve translation invariance.

(2) **Periodic boundary conditions:** Such an approach satisfies translation invariance. Consider for example a square lattice that consists of 7×7 sites as shown in Fig. 6.4 (the following discussion can also be generalized to more complicated lattices). Periodic boundary conditions imply that sites outside the array can be identified with sites located inside. For the following consideration we marked a reference site i_1 in the center of the array. This site is of course arbitrary. Note that in such a system the distance of a two-particle vertex is limited to a maximum of 3 lattice constants in each direction. Longer ones can be identified with vertices within this limitation. The site summation in the first flow equation, see Fig. 6.3a, is simply performed by keeping the reference site i_1 fixed and adding up all vertices $\Gamma^\Lambda_{\text{d}\, i_1 i_2}$ when i_2 runs over the 7×7 site-array. The other site summation in the RPA term of the second flow equation (see Fig. 6.2b or the sum term in Eq. (6.45)) is handled as follows: Imagine that on the left hand side of the equation we consider the flow of the vertex with site indices i_1 and i_2 (dotted line in Fig. 6.4). The RPA term exhibits a site summation (with index j) over two vertices, the first one connecting i_1 and j and the second one connecting j and i_2. Then the sum $\sum_j$ runs over all 7×7 sites. There are however sites j for which the vertex between j and i_2 seems to be longer than the maximum of 3 lattice constants in each direction (those sites outside the dashed box in Fig. 6.4) but due to periodic boundary conditions such a vertex can be related to a vertex within this limitation. Compared to the hard-wall boundary conditions this scheme has the advantage that translation invariance is guaranteed and that surface effects do not occur. On the other hand it has the disadvantage that it still treats a finite system and that the site summation $\sum_j$ is rather large. The scheme presented next leads to an improvement in these aspects.

(3) **Infinite system with spatially limited vertices:** This approach assumes an infinitely large system but two-particle vertices are only taken into account if their distance vector $\mathbf{R} = \mathbf{R}_{i_1} - \mathbf{R}_{i_2}$ does not exceed a maximal length in each direction.

In other words **R** has to satisfy the condition

$$\text{Max}(|R_x|, |R_y|) \leq L \tag{6.47}$$

with a given integer L. Longer vertices are treated as zero for all Λ. If L is chosen as $L = 3$ then the set of the included vertices is formed by the bonds from the center site i_1 to all other sites of the array in Fig. 6.4. The sum in the first flow equation is performed as for periodic boundary conditions: The site i_1 is kept fixed while i_2 runs over all 7×7 surrounding sites. In contrast, the sum over j in the RPA term (assuming again that the flow of the vertex $\Gamma^\Lambda_{\text{s/d}\, i_1 i_2}$ is considered) only runs over the sites inside the dashed box because only for these sites the condition $\text{Max}(|R_x|, |R_y|) \leq 3$ is satisfied for both, the vertex between i_1 and j as well as the vertex between j and i_2. Consequently, compared to periodic boundary conditions this scheme has the advantage that the summation range of $\sum_j$ is reduced, speeding up the numerics. Furthermore, the system size within this scheme is in principle infinitely large. The condition (6.47) is of course an approximation for the spatial dependence of the two-particle vertex but not in the sense that it implements a finite system. In all our numerics we use this approach. However, in order to have a measure for the system size (for example for comparison to other methods considering finite systems) one may relate a system with a given L to a lattice consisting of $(2L + 1)^2$ sites, due to the similarities to periodic boundary conditions.

Treatment of the Frequency Dependences

Having discussed the spatial properties of the flow equations in detail we now turn to the frequency dependences. A first glance at Eqs. (B.1) and (B.2) reveals a weird frequency structure but some important features are visible. Each two-particle vertex Γ^Λ_s or Γ^Λ_d appearing on the right hand side has exactly one of the invariant frequencies s, t, u in its arguments. From this one may distinguish the different channels: The particle-particle channel belongs to the frequency s, the direct particle-hole channel to t and the crossed particle-hole channel to u. The quantities s, t, u, are the frequencies running along the internal fermion bubbles formed by G^Λ and S^Λ. Hence, they are also referred to as transfer energies. The use of s, t, u as arguments has the advantage that the vertex $\Gamma^\Lambda_{\text{s/d}\, i_1 i_2}(s, t, u)$ is invariant under each of the transformations $s \to -s$, $t \to -t$ and $u \to -u$. Additionally $\Gamma^\Lambda_{s\, i_1 i_2}(s, t, u)$ is invariant under $s \leftrightarrow u$ and $\Gamma^\Lambda_{d\, i_1 i_2}(s, t, u)$ changes sign under $s \leftrightarrow u$. These four symmetries reduce the numerical computation time by a factor of $2^4 = 16$ because only $1/16$ of the space spanned by s, t, u needs to be considered. The use of these symmetries is essential in order to keep the computation times tolerable. They are proved in appendix C.

For our numerics we need to approximate the frequency dependences. A finite set of equations is obtained only if all continuous frequency arguments are discretized. Typically we use a combination of a linear and a logarithmic mesh symmetrically arranged around zero frequency (the zero-frequency component itself is not included in the mesh because vertex functions might be non-analytic there). The linear behavior applies to the highest frequencies of the mesh while the logarithmic part assures that the physics at small energy scales is sufficiently resolved. For all frequency arguments of γ^Λ and $\Gamma^\Lambda_{s/d}$ the same mesh is used. In general, regarding the first FRG equation, Eq. (6.40), for each discrete frequency ω one flow equation is obtained, similarly applying to the second FRG equation: Each set of discrete values s, t, u and each distance vector $\mathbf{R}_{i_1} - \mathbf{R}_{i_2}$ is associated with one differential equation for $\Gamma^\Lambda_{s\,i_1 i_2}(s,t,u)$ and $\Gamma^\Lambda_{d\,i_1 i_2}(s,t,u)$. Because of the lattice sums and the irregular structure of the frequency arguments these equations are highly coupled to each other.

In the general case the frequency arguments ω, ω_1, ω_2, ω_3 appearing in $\gamma^\Lambda(\omega)$ and $\Gamma^\Lambda_{s/d\,i_1 i_2}(\omega_1, \omega_2, \omega_3)$ on the right hand sides of the flow equations do not coincide with discrete mesh points (however, as already pointed out, one of the frequencies ω_1, ω_2, ω_3 is always given by an external frequency s, t or u and is therefore part of the mesh). Hence, we need a scheme that associates arbitrary $\gamma^\Lambda(\omega)$ and $\Gamma^\Lambda_{s/d\,i_1 i_2}(\omega_1, \omega_2, \omega_3)$ with components $\gamma^\Lambda(\omega_n)$ and $\Gamma^\Lambda_{s/d\,i_1 i_2}(\omega_{n_1}, \omega_{n_2}, \omega_{n_3})$, respectively, where the frequencies with an index $n_{(i)}$ are understood as discrete mesh values. The simplest method is to choose the nearest discrete mesh point in the frequency space. However, since frequency arguments typically change continuously during the Λ-flow, the nearest mesh point will often jump to another value leading to an uncontrolled flow behavior. We do not employ this scheme in our numerics. Instead we use a linear interpolation between mesh points, that is, for an arbitrary frequency ω the self energy $\gamma^\Lambda(\omega)$ is evaluated as

$$\gamma^\Lambda(\omega) = \left[\gamma^\Lambda(\omega_<)(\omega_> - \omega) + \gamma^\Lambda(\omega_>)(\omega - \omega_<)\right] \frac{1}{\omega_> - \omega_<}, \tag{6.48}$$

where $\omega_<$ is the nearest mesh point smaller than ω and $\omega_>$ is the nearest mesh point larger than ω. This scheme is straightforwardly extended to higher dimensions. For example, a vertex of the form $\Gamma^\Lambda_{s/d\,i_1 i_2}(\omega_1, \omega_2, u)$ as it appears in the crossed particle-hole channel is approximated by

$$\begin{aligned}
\Gamma^\Lambda_{s/d\,i_1 i_2}(\omega_1, \omega_2, u) = \big[\; & \Gamma^\Lambda_{s/d\,i_1 i_2}(\omega_{1<}, \omega_{2<}, u)(\omega_{1>} - \omega_1)(\omega_{2>} - \omega_2) \\
& + \Gamma^\Lambda_{s/d\,i_1 i_2}(\omega_{1<}, \omega_{2>}, u)(\omega_{1>} - \omega_1)(\omega_2 - \omega_{2<}) \\
& + \Gamma^\Lambda_{s/d\,i_1 i_2}(\omega_{1>}, \omega_{2<}, u)(\omega_1 - \omega_{1<})(\omega_{2>} - \omega_2) \\
& + \Gamma^\Lambda_{s/d\,i_1 i_2}(\omega_{1>}, \omega_{2>}, u)(\omega_1 - \omega_{1<})(\omega_2 - \omega_{2<}) \big] \\
& \times \frac{1}{(\omega_{1>} - \omega_{1<})(\omega_{2>} - \omega_{2<})}.
\end{aligned} \tag{6.49}$$

Here u is assumed to coincide with a mesh point. Despite the numerical effort associated with the evaluation of such an expression it is worth to implement this scheme since it smooths the flow considerably.

Numerical Solution of the Flow Equation
The solutions of the equations (6.40), (B.1) and (B.2) at a given Λ_0 are found by integrating both sides,

$$\gamma^{\Lambda_0}(\omega) = \gamma^{\Lambda \to \infty} - \int_{\Lambda_0}^{\infty} d\Lambda \,[\text{r.h.s. of Eq. (6.40)}],$$

$$\Gamma_{\text{s/d }i_1 i_2}^{\Lambda_0}(s,t,u) = \Gamma_{\text{s/d }i_1 i_2}^{\Lambda \to \infty}(s,t,u) - \int_{\Lambda_0}^{\infty} d\Lambda \,[\text{r.h.s. of Eq. (B.1)/(B.2)}]. \quad (6.50)$$

It turned out that the Euler method for solving ordinary differential equations is sufficient, i.e., in order to evaluate the integrals numerically, the integration range $[\Lambda_0, \infty]$ is split up into small intervals $[\Lambda_0, \Lambda_1]$, $[\Lambda_1, \Lambda_2]$, ..., $[\Lambda_{m-1}, \Lambda_m]$ and the boundary $\Lambda = \infty$ is replaced by some large but finite value Λ_m. Each interval $[\Lambda_n, \Lambda_{n+1}]$ constitutes one RG step which yields numerical values for γ^{Λ_n} and $\Gamma_{\text{s/d}}^{\Lambda_n}$. For the integration $\int_{\Lambda_n}^{\Lambda_{n+1}} d\Lambda \ldots$ being performed in this step all self energies and two-particle vertices appearing in the integrand are taken from the preceding RG step, i.e., $\gamma^{\Lambda_{n-1}}$ and $\Gamma_{\text{s/d}}^{\Lambda_{n-1}}$, and are assumed to be constant in Λ within the interval. The remaining explicit Λ-dependence of the integrand may be treated by analytical integration. In this way the equations are solved for arbitrary energy scales Λ_0. Of special interest is of course the cutoff-free case $\Lambda_0 = 0$ where the physical model is recovered. However, for the numerical solution the step size needs to be reduced during the flow such that the limit $\Lambda = 0$ is never reached exactly. We typically stop the flow at scales several magnitudes below the coupling strength, i.e., in an energy regime where no relevant change in the physics is expected.

Spin-Correlation Function and Susceptibility
Primarily, we are less interested in the vertex functions γ^Λ and $\Gamma_{\text{s/d}}^\Lambda$ but rather in physical observables. A quantity that measures the magnetic properties and that is accessible within our approach is the spin-correlation function $\chi_{ij}(i\nu)$ defined by

$$\chi_{ij}(i\nu) = \int_0^\infty d\tau e^{i\nu\tau} \langle T_\tau \{S_i^z(\tau) S_j^z(0)\} \rangle. \quad (6.51)$$

Expressing the spins in terms of pseudo fermions and expanding the expectation value diagrammatically, this object may be related to vertex functions as follows,

$$\chi_{ij}^\Lambda = \langle i \quad j \rangle \delta_{ij} + \langle i \quad j \rangle. \quad (6.52)$$

In explicit form this equation reads,

$$\chi_{ij}^\Lambda(i\nu) = -\frac{1}{2}\frac{1}{2\pi}\int d\omega \, G^\Lambda(\omega) G^\Lambda(\omega+\nu) \, \delta_{ij}$$

$$-\frac{1}{4}\left(\frac{1}{2\pi}\right)^2 \int\int d\omega d\omega' \, G^\Lambda(\omega) G^\Lambda(\omega+\nu) G^\Lambda(\omega') G^\Lambda(\omega'+\nu)$$

$$\times \sum_{\alpha_{1'}\alpha_{2'}\alpha_1\alpha_2} \Gamma^\Lambda(1',2';1,2) \sigma_{\alpha_1\alpha_{1'}}^z \sigma_{\alpha_2\alpha_{2'}}^z, \quad (6.53)$$

where we already generalized $\chi_{ij}(i\nu)$ to $\chi_{ij}^\Lambda(i\nu)$ by decorating each flowing object with an index Λ. The propagators G^Λ are the dressed ones defined in Eq. (6.29). Remind the notation $1 = \{\omega_1, i_1, \alpha_1\}$. In Eq. (6.53) we have $i_1 = i_{1'} = i$, $i_2 = i_{2'} = j$ and $\omega_{1'} = \omega + \nu$, $\omega_{2'} = \omega'$, $\omega_1 = \omega$, $\omega_2 = \omega' + \nu$. Since rotational invariance is assumed, the direction of the spin correlations can be chosen arbitrarily (in our case the z-direction). Inserting the parametrization (6.36) and using the transfer frequencies, we find

$$\chi_{ij}^\Lambda(i\nu) = -\frac{1}{4\pi} \int d\omega\, G^\Lambda(\omega) G^\Lambda(\omega+\nu)\, \delta_{ij}$$
$$-\frac{1}{8\pi^2} \int\int d\omega\, d\omega'\, G^\Lambda(\omega) G^\Lambda(\omega+\nu) G^\Lambda(\omega') G^\Lambda(\omega'+\nu)$$
$$\times \left[2\Gamma_{s\,ij}^\Lambda(\omega+\omega'+\nu, \nu, \omega-\omega') + \Gamma_{s\,ii}^\Lambda(\omega+\omega'+\nu, \omega-\omega', \nu)\delta_{ij} \right.$$
$$\left. -\Gamma_{d\,ii}^\Lambda(\omega+\omega'+\nu, \omega-\omega', \nu)\delta_{ij} \right]. \qquad (6.54)$$

In each RG step we evaluate this equation for all possible distances $\mathbf{R}_i - \mathbf{R}_j$. The frequency integrations have to be performed numerically. Primarily, we are interested in the static correlation function $\chi_{ij}^\Lambda(i\nu = 0)$ but in principle the dynamics is accessible as well. A quantity measurable in experiment is the magnetic susceptibility $\chi(\mathbf{k}, i\nu)$ which is defined in analogy to Eq. (6.51),

$$\chi(\mathbf{k}, i\nu) = \int_0^\infty d\tau e^{i\nu\tau} \langle T_\tau \{ S^z(-\mathbf{k}, \tau) S^z(\mathbf{k}, 0)\} \rangle, \qquad (6.55)$$

with

$$S^z(\mathbf{k}, \tau) = \frac{1}{\sqrt{N}} \sum_i e^{-i\mathbf{k}\mathbf{R}_i} S_i^z. \qquad (6.56)$$

Inserting Eq. (6.56) into Eq. (6.55) (and adding the label Λ for flowing quantities) we obtain

$$\chi^\Lambda(\mathbf{k}, i\nu) = \sum_j e^{i\mathbf{k}(\mathbf{R}_i - \mathbf{R}_j)} \chi_{ij}^\Lambda(i\nu), \qquad (6.57)$$

where (at least for lattices with a monoatomic unit cell) the site i can be chosen arbitrarily. That is, the $\mathbf{k}$-space resolved susceptibility is the Fourier transform of the correlation function in real space. Typically, for a given spin model we study the behavior of the susceptibility $\chi^\Lambda(\mathbf{k})$ for different wave vectors $\mathbf{k}$ during the Λ-flow.

6.3 Static FRG Approach

The numerical treatment of the frequency dependences requires a lot of computational effort. Hence, it is tempting a use a scheme that neglects these dependences. Before we consider the general case including all frequencies we briefly discuss the feasibility of a static approximation which takes into account only the zero-frequency components of the vertex functions. Such a scheme leads to a huge reduction of the complexity of the flow equations. However, putting all frequency arguments in the first flow equation (6.40) to zero, the right hand side vanishes and the self energy remains zero during the flow. This can be traced back to the self energy being an odd function in the frequency. As stated in Chapter 4, the effect of frustration in destroying magnetic order cannot be described without the influence of the self energy. Therefore, in order to allow for a broadening of the spectrum we again assume the discontinuous form $\Sigma^\Lambda = -i\gamma^\Lambda \, \text{sgn}(\omega)$, see Eq. (5.5). However, inserting this form together with the static two-particle vertex into the first flow equation (6.40) leads again to a vanishing flow, $\frac{d\gamma^\Lambda}{d\Lambda} \equiv 0$. It is not possible to obtain a finite flow for γ^Λ within a static scheme. Hence, γ^Λ has to be considered again as a given phenomenological parameter that is independent of Λ.

We emphasize that in some way the truncation of the flow equations gets simpler within such a scheme: The conventional truncation and the Katanin truncation discussed in the next two sections are identical because the extra term in the single-scale propagator generated by the Katanin scheme contains the derivative of the self energy, $\frac{d}{d\Lambda}\gamma^\Lambda$, which vanishes here. Since an explicit flow of the three-particle vertex in not considered, the two-particle vertex is the only flowing quantity.

Flow equations for the zero-frequency components of $\Gamma^\Lambda_{s\,i_1 i_2}$ and $\Gamma^\Lambda_{d\,i_1 i_2}$ are obtained by putting s, t and u equal to zero in Eqs. (B.1) and (B.2). Thereafter, the right hand sides still contain finite-frequency components which are approximated by the static ones. Note that the frequency dependence of $\Sigma = -i\gamma\,\text{sgn}(\omega)$ only affects the internal integration. We obtain static flow equations of the form

$$\frac{d}{d\Lambda}\Gamma^\Lambda_{s\,i_1 i_2} = \frac{2}{\pi}\frac{1}{(\Lambda+\gamma)^2}\left[\sum_j \Gamma^\Lambda_{s\,i_1 j}\Gamma^\Lambda_{s\,j i_2} - 2\left(\Gamma^\Lambda_{s\,i_1 i_2}\right)^2 + \Gamma^\Lambda_{s\,i_1 i_2}\left(\Gamma^\Lambda_{s\,i_1 i_1} - \Gamma^\Lambda_{d\,i_1 i_1}\right)\right], \quad (6.58a)$$

$$\frac{d}{d\Lambda}\Gamma^\Lambda_{d\,i_1 i_2} = \frac{2}{\pi}\frac{1}{(\Lambda+\gamma)^2}\left[\sum_j \Gamma^\Lambda_{d\,i_1 j}\Gamma^\Lambda_{d\,j i_2} - \Gamma^\Lambda_{d\,i_1 i_2}\left(3\Gamma^\Lambda_{s\,i_1 i_1} + \Gamma^\Lambda_{d\,i_1 i_1}\right)\right], \quad (6.58b)$$

with the initial conditions

$$\Gamma^{\Lambda\to\infty}_{s\,i_1 i_2} = \frac{1}{4}J_{i_1 i_2}, \quad (6.59a)$$

$$\Gamma^{\Lambda\to\infty}_{d\,i_1 i_2} = 0. \quad (6.59b)$$

Obviously, Eqs. (6.58b) and (6.59b) are trivially fulfilled by $\Gamma^\Lambda_{d\,i_1 i_2} \equiv 0$, such that only Eq. (6.58a) has to be solved numerically. The spatial dependence of this equation is treated as described in the previous section, i.e., vertices exceeding a certain distance are neglected. A flow towards finite values for $\Lambda \to 0$ indicates a paramagnetic phase while a diverging flow is a sign of a magnetic instability. The type of order can be extracted by transforming $\Gamma^\Lambda_{s\,i_1 i_2}$ into Fourier-space and identifying the fastest momentum component. Applying such a scheme to the J_1-J_2 model one can draw a phase diagram in the γ-g-plane, see Fig. 6.5. The figure compares the phase boundaries with the results from the finite-lifetime theory in Chapter 5. Note that both approaches are closely related because the FRG includes an RPA contribution in the form of the diagram in Fig. 6.2b, representing the direct particle-particle channel. Correspondingly, the two phase diagrams look rather similarly. Again the boundaries are given by straight lines,

6 The Functional Renormalization Group: Implementation for Spin Systems

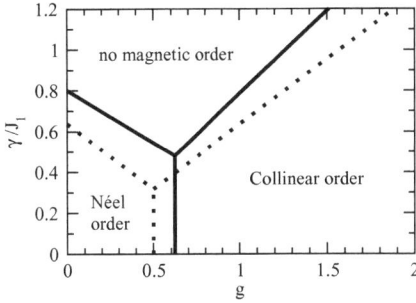

Figure 6.5: Phase diagram in the γ-g-plane for a static FRG approximation with a phenomenological parameter γ (full line). The dotted line shows the phase boundaries of the finite-lifetime ansatz from Fig. 5.2.

but the value of the frustration g for which the paramagnetic phase has its largest extent moved from $g = 0.5$ to $g \approx 0.62$. This demonstrates that the RPA term is the dominant one because it is responsible for the formation of magnetic order. On the other hand, the remaining contributions, i.e., Fig. 6.2a, 6.2c, 6.2d and 6.2e, are only corrections that do not modify the phase diagram qualitatively.

If we neglect these remaining terms, we can in fact show that the RPA results of the finite-lifetime approach of Chapter 5 are reproduced. Singling out the direct particle-hole channel, i.e., the sum term in Eq. (6.58a) we obtain

$$\frac{d}{d\Lambda}\Gamma^\Lambda_{s\,i_1 i_2} = \frac{2}{\pi}\frac{1}{(\Lambda+\gamma)^2}\sum_j \Gamma^\Lambda_{s\,i_1 j}\Gamma^\Lambda_{s\,j i_2}\,. \qquad (6.60)$$

With the numerical solution for $\Gamma^\Lambda_{s\,i_1 i_2}$, the correlation function $\chi^\Lambda_{ij}(i\nu = 0)$ can be computed using Eq. (6.54) which in our case simplifies to

$$\chi^\Lambda_{ij}(i\nu = 0) = \frac{1}{2\pi(\Lambda+\gamma)}\delta_{ij} - \frac{1}{\pi^2}\frac{1}{(\Lambda+\gamma)^2}\Gamma^\Lambda_{s\,ij}\,. \qquad (6.61)$$

In fact, if the RPA term is considered individually, as in Eq. (6.60), the restriction to static vertices is not an approximation because the zero-frequency component does not couple to finite frequencies. The results are shown in Fig. 6.6 where in accordance with Fig. 5.3 the damping parameter is chosen as $\gamma = 0.36 J_1\sqrt{1+g^2}$. In the ordered regimes $0 \leq g \lesssim 0.39$ and $0.69 \lesssim g$ the corresponding ordering-vector component of the susceptibility runs into a divergence while in the intermediate paramagnetic regime finite values are obtained in the limit $\Lambda \to 0$. It is also shown that these endpoints indeed coincide with the RPA-susceptibility data from Fig. 5.3.

As an application of this scheme in a different context, exact mean-field results for the reduced BCS model of superconductors have been reproduced [107]. In particular, in conjunction with a small symmetry-breaking external field it has been shown that symmetry-broken phases are accessible (see Refs. [51, 52, 107] and Chapter 8).

In summary, the considerations in this section led to the conclusion that a static approximation of the FRG equations does not allow to calculate the central quantity

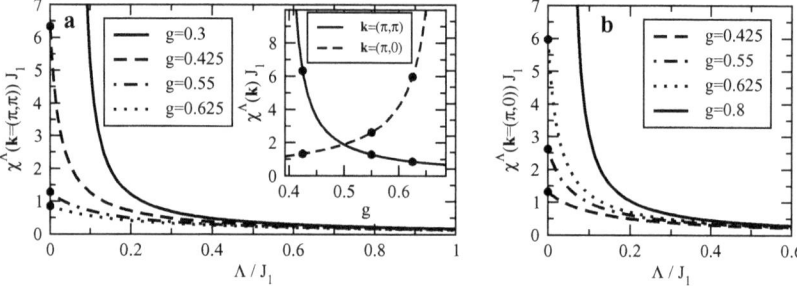

Figure 6.6: Flowing susceptibility for an static FRG scheme in the RPA channel. (a) displays the Néel susceptibility and (b) the collinear susceptibility. The phenomenological damping parameter is chosen as $\gamma = 0.36 J_1 \sqrt{1+g^2}$, with different values for g. The inset shows the susceptibilities of Fig. 5.3 together with the endpoints of the flow taken from the two main figures (circles). It is seen that both approaches yield the same results.

governing the destruction of long-range order: the pseudo-fermion spectral width γ. On the basis of a static FRG approach that treats the damping as a phenomenological input, we find that the RPA term is the dominant contribution while the other terms are corrections. Moreover, we have shown that within an FRG scheme in the direct particle-hole channel the RPA results of Chapter 5 are reproduced. In general, it is rather difficult to neglect frequency dependences but keeping the effect on the solution small. Thus, in the remaining part of this thesis we treat the FRG equations in full complexity, including the dynamics of all frequency dependences. This will lead to a finite spectral broadening without further assumptions. However, once the damping γ is considered as a flowing quantity, the truncation scheme requires a closer inspection, see the discussion in the next two sections.

6.4 Conventional Truncation Scheme

While the approximations concerning the frequency discretization, the spatial truncation or the finite RG step size, as discussed in Section 6.2 are relatively easy to handle, the truncation of the FRG equations poses a severe problem. Most FRG studies truncate the hierarchy by completely neglecting the three-particle contribution. For models with small or moderate interaction strength this scheme may be well controlled because the three-particle vertex is at least of third order in the couplings. However, in the strong coupling limit considered in this thesis it might be insufficient. Nevertheless, in a first test of our implementation we apply this scheme, which we refer to as the conventional

6 The Functional Renormalization Group: Implementation for Spin Systems

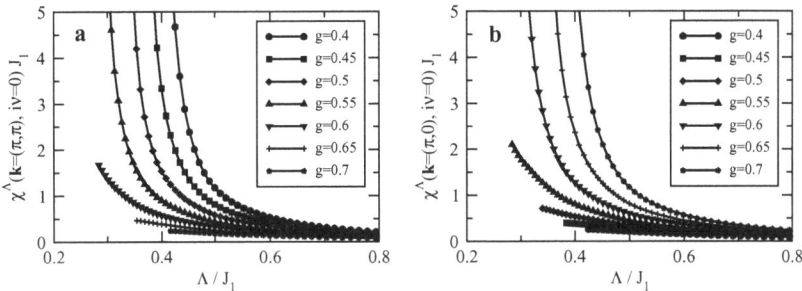

Figure 6.7: Flow of the static Néel susceptibility (a) and collinear susceptibility (b) for different frustrations g within the conventional truncation scheme.

truncation scheme.

Within this approach, the single-scale propagator is defined in the usual way, see Eq. (6.32). Due to its δ-singularity, the internal frequency integrations in the corresponding flow equations (6.40), (B.1) and (B.2) can be performed analytically, $\int_{-\infty}^{\infty} d\omega' \delta(|\omega'| - \Lambda) \ldots = \sum_{\omega' = \pm \Lambda} \ldots$, simplifying the numerics considerably. In particular, in order to account for the influence of a spectral broadening it is essential to include the feedback of the self energy into the flow of the two-particle vertex. The flow equations (6.40), (B.1) and (B.2) are solved for the J_1-J_2 model using different frustration parameters $g = \frac{J_2}{J_1}$ as described in Section 6.2. In each RG step the magnetic susceptibilities with momenta $\mathbf{k} = (\pi, \pi)$ and $\mathbf{k} = (\pi, 0)$ are calculated, corresponding to Néel order and collinear order, respectively. Since the two-particle vertex depends on three frequencies the computational effort grows with the third power of the number of discrete frequencies. Regarding the truncation in real space the computing time grows with the fourth power of the linear dimension L, see Eq. (6.47), because two spatial dimensions lead to a quadratic dependence and the internal site summation provides two extra powers.

The results are plotted in Fig. 6.7. It is seen that during the flow the Néel susceptibility exhibits a divergence for all $g \lesssim 0.55$. On the other hand the collinear susceptibility appears to diverge for all $g \gtrsim 0.55$. In particular, there is no parameter region without a magnetic instability such that the flow never reaches $\Lambda = 0$. In the present approximation the paramagnetic phase is obviously missing. The abrupt stop of the flow of the susceptibilities for $g \geq 0.6$ in Fig. 6.7a and $g \leq 0.55$ in Fig. 6.7b can be traced to the divergence of the respective other susceptibility. For g at the transition, i.e., between 0.55 and 0.6, the divergence is clearly indicated at the smallest accessible Λ.

The absence of an intermediate non-magnetic phase is quite unsatisfactory. In general, the flow behavior of the two-particle vertex is strongly influenced by the term $\frac{1}{\Lambda + \gamma^\Lambda(\Lambda)}$ which is part of the internal fermion bubble P^Λ in the differential equations (B.1) and (B.2). As long as the self energy is not too large, this term increases with decreasing

Λ and generates the strong rise of the leading susceptibility component. Since the two-particle vertex couples to the first flow equation, also the self energy undergoes a strong increase. Hence, on an approach to the instabilities seen in Fig. 6.7, we obtain a diverging solution of both, the two-particle vertex and the self energy. This behavior is remarkable: Although a large self energy regularizes the pole of $\frac{1}{\Lambda+\gamma^\Lambda(\Lambda)}$ the feedback of the self energy into the two-particle vertex described by that term seems to be too weak to stop the divergence of the two-particle vertex. In other words, the mechanism of Chapter 5 demonstrating that a large self energy, i.e., pseudo-fermion damping generates disorder fluctuations obviously fails here. This can be traced back to the fact that within the conventional truncation scheme the dressed RPA (i.e., the RPA including renormalized Green's functions) is not recovered (see next section). We emphasize that on the other hand, within an approximation scheme that includes the dressed RPA and hence ensures a proper feedback of the self energy into the two-particle vertex, the behavior observed here cannot occur: A large self energy would generate strong damping effects which stop the divergence of the susceptibility, as shown in Chapter 5. An improvement of such kind will be discussed in the next section. The drawback reported here is actually of general type: Even for a model consisting of two coupled spins, $H = J\mathbf{S}_1\mathbf{S}_2$, the correlation function χ_{12}^Λ between the sites diverges during the flow, demonstrating that magnetic instabilities occur systematically within this scheme. Indeed, in some way, the conventional truncation reproduces the solution of a bare RPA scheme which generates magnetic order regardless of the dimension.

In summary, the conventional truncation scheme proved to be insufficient to describe the melting of long-range order as a result of quantum fluctuations. We identify this drawback as a consequence of an insufficient feedback of the self energy into the two-particle vertex. In fact, it turns out that this feedback is so weak that our solutions resemble the results of a bare RPA scheme where the self energy is completely neglected. An improvement is discussed in the next section where we consider a modified truncation scheme.

6.5 Katanin Truncation Scheme

In general, truncating the infinite hierarchy of FRG flow equations is not consistent with the fulfillment of conservation laws, expressed in terms of Ward identities. For the conventional one-loop truncation scheme it can be shown [67] that Ward identities are only fulfilled up to terms of third power in the two-particle vertex. Guided by the idea of an improved preservation of Ward identities, Katanin recently formulated a modified scheme [67]. In its original version this approach is based on the two-loop FRG equations, which include the flow of the three-particle vertex. For these three-particle contributions one distinguishes between vertex-correction terms and self-energy correction terms, see [67]. Modifying the flow of the two-loop self-energy corrections in a certain way it is shown that the fulfillment of Ward identities is improved in the sense that they are violated only by terms with overlapping loops of fourth order in the two-particle vertex (instead of a violation by overlapping and non-overlapping loop terms in the

6 The Functional Renormalization Group: Implementation for Spin Systems

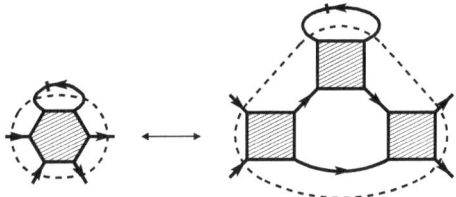

Figure 6.8: Three-particle contribution to the two-particle vertex flow (left) and an example for an extra term generated within the Katanin truncation (right). The diagram on the right is constructed by replacing the single-scale propagator in the particle-particle channel by a self-energy correction. Similar diagrams exist for the other channels. Comparison shows that both terms are of the same type (the three-particle vertices are marked with dashed lines).

unmodified two-loop equations). As a matter of fact, due to the two-loop contributions, the numerical solution of such a scheme is very challenging. Thus, the full Katanin regularization has again been modified such that it became amenable to applications. It is this second version that is now known as the Katanin truncation. Despite its simplifications it still leads to essential improvements compared to the conventional truncation, as demonstrated below.

Starting from the one-loop equation (6.26b) the basic change is the substitution of the single-scale propagator as it is defined in Eq. (6.30) by the total derivative of $-G^\Lambda$ with respect to Λ,

$$S^\Lambda(\omega) \to -\frac{d}{d\Lambda} G^\Lambda(\omega) = S^\Lambda(\omega) - \left(G^\Lambda(\omega)\right)^2 \frac{d}{d\Lambda} \Sigma^\Lambda(\omega). \tag{6.62}$$

In the expression on the right side the first term stems from the derivative of the cutoff function and reproduces the standard single-scale propagator while the second term takes into account the flow of the self energy. We emphasize that the replacement (6.62) is only made in the second flow equation (6.26b) but not in the first flow equation (6.26a). In diagrammatic representation the substitution may be displayed as

$$\longrightarrow \quad \longrightarrow \quad -\frac{d}{d\Lambda} \longrightarrow \quad = \quad \longrightarrow \quad - \quad \longrightarrow \quad , \tag{6.63}$$

where the first flow equation (6.26a) has been used for the derivative of the self energy. The second contribution on right side of Eq. (6.63) generates extra terms if inserted in the flow equation for the two-particle vertex. An example of such an additional term is presented in Fig. 6.8. Comparison to the previously neglected diagram shows that the extra terms indeed contribute to the three-particle vertex flow. These special three-particle terms play a crucial role since they assure that the dressed RPA resummation is recovered as a diagram subset. This issue will be exemplified below.

Katanin Truncation and RPA

As already indicated in the previous section, the property of the Katanin truncation-scheme to recover the dressed RPA [107] leads to a substantial improvement compared to the conventional one-loop truncation: It ensures that the coupling between the self-energy flow and the two-particle vertex flow is strong enough such that the self energy is able to destroy magnetic order (however, it does not yet ensure that the damping indeed has the right size to generate the correct phase diagram, see the discussion below). In the following we discuss this property of the Katanin truncation in some detail. As pointed out earlier, the RPA contribution is generated by the direct particle-hole channel, i.e., by the diagram in Fig. 6.2b containing a site summation. Singling out this term in the flow equation (B.6), we are left with

$$\frac{d}{d\Lambda}\Gamma^\Lambda_{s\,i_1 i_2}(s,t,u) = \frac{1}{\pi}\int_{-\infty}^{\infty} d\omega' \sum_j \Gamma^\Lambda_{s\,i_1 j}(\omega_{1'}+\omega', t, \omega_1-\omega')\Gamma^\Lambda_{s\,ji_2}(\omega_2+\omega', t, -\omega_{2'}+\omega')$$
$$\times \left[P^\Lambda(\omega', t+\omega') + P^\Lambda(t+\omega', \omega')\right], \qquad (6.64)$$

where P^Λ denotes the internal fermion loop consisting of the single-scale propagator (as it is defined within the Katanin scheme) and the Green's function,

$$P^\Lambda(\omega_1, \omega_2) = -S^\Lambda(\omega_1)G^\Lambda(\omega_2) = \left[\frac{d}{d\Lambda}G^\Lambda(\omega_1)\right]G^\Lambda(\omega_2), \qquad (6.65)$$

see also Eq. (B.5). Note that the RPA contribution in the density-interaction sector, Eq. (B.7), does not develop a finite flow because it is zero in the initial conditions. The right hand side of Eq. (6.64) includes the transfer frequencies s and u only in the arguments of Γ_s. As long as the initial conditions for Γ_s are independent of s and u, these frequency dependencies may also be omitted at finite Λ. On the other hand, the variable t explicitly appears in the fermion loop. Consequently, Eq. (6.64) can be written as

$$\frac{d}{d\Lambda}\Gamma^\Lambda_{s\,i_1 i_2}(t) = \frac{1}{\pi}\sum_j \Gamma^\Lambda_{s\,i_1 j}(t)\Gamma^\Lambda_{s\,ji_2}(t)\int_{-\infty}^{\infty} d\omega' \frac{d}{d\Lambda}\left[G^\Lambda(\omega')G^\Lambda(t+\omega')\right]. \qquad (6.66)$$

This equation has a unique solution which satisfies

$$\Gamma^\Lambda_{s\,i_1 i_2}(t) = \Gamma^{\Lambda\to\infty}_{s\,i_1 i_2} + \frac{1}{\pi}\sum_j \Gamma^{\Lambda\to\infty}_{s\,i_1 j}\Gamma^\Lambda_{s\,ji_2}(t)\int_{-\infty}^{\infty} d\omega' G^\Lambda(\omega')G^\Lambda(t+\omega'). \qquad (6.67)$$

Due to $G^{\Lambda\to\infty}(\omega) = 0$ the initial conditions are correctly incorporated in this equation. In order to motivate this identity we show a diagrammatic proof in Fig. 6.9. In principle, Eq. (6.67) can be traced back to the fact that the Katanin truncation includes a total derivative of the fermion bubble as indicated in Eq. (6.66). Indeed, Eq. (6.67) or Fig. 6.9a

6 The Functional Renormalization Group: Implementation for Spin Systems

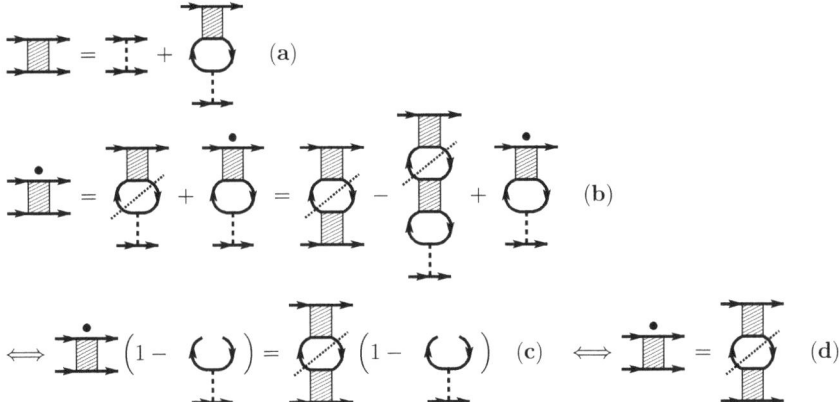

Figure 6.9: Diagrammatic proof of Eq. (6.66) using the identity (6.67). Dots and dotted lines indicate Λ-derivatives. Note that the dotted lines denote total derivatives of the dressed fermion loop. The equation (a) corresponds to Eq. (6.67). (b) represents the derivative of (a). The right hand side of (b) is obtained by inserting the bare vertex in (a) into the first graph of the middle part in (b). The equation (b) is equivalent to equation (c) which directly leads to Eq. (6.66) and to its diagrammatic analog (d).

are representations of the RPA bubble-chain which proves that the Katanin truncation reproduces the dressed RPA.

Moreover, using Eq. (6.67), it can be shown that on the single-particle level the Hartree self-energy is recovered (if it is finite). In Fig. 6.10 we present the corresponding diagrammatic proof. Together with Fig. 6.9 it is now obvious that the full RPA + Hartree resummation is included in the Katanin scheme. Within the present SU(2)-invariant formulation, however, the Hartree self-energy vanishes. In Chapter 8, where we explicitly take into account symmetry breaking in the form of a small magnetic field in the initial conditions, the Hartree term is finite and plays a crucial role when the flow approaches ordered phases.

The above considerations may be straightforwardly extended to other interaction channels. In particular, it can be shown that the full (dressed) particle-particle ladder as well as the crossed particle-hole ladder contribute to the approximation. On the level of self energies the crossed particle-hole ladder ensures the inclusion of the Fock graph while the particle-particle ladder generates a Fock self-energy with an anomalous propagator. However, both self-energy contributions vanish because the pseudo-fermion constraint forbids hopping processes. Nonetheless, the inclusion of the ladder diagrams is essential in our scheme. We emphasize that the RPA bubble-chain forms the leading contribution in a $\frac{1}{S}$ expansion, where S is the spin length. In a classical system with $S \to \infty$ the bare RPA becomes exact such that paramagnetic phases are suppressed. Hence, the RPA is

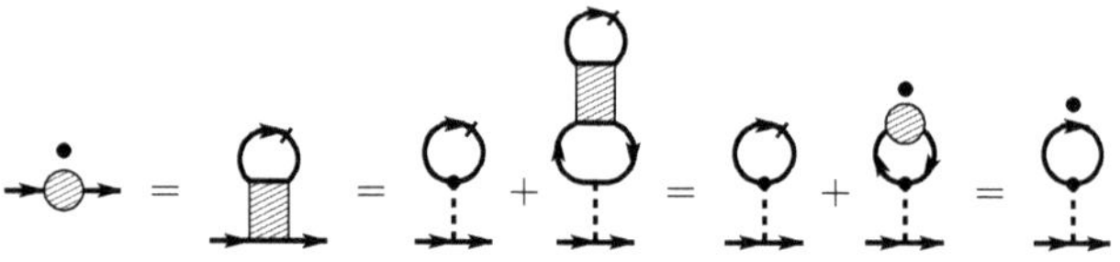

6 The Functional Renormalization Group: Implementation for Spin Systems

Figure 6.10: The Hartree resummation within the Katanin scheme. Note that the single-scale propagator in its conventional form, see Eq. (6.30), is given by a slashed line whereas total derivatives are represented by dots. The first equation is the FRG flow equation for the self energy. In the second step the right hand side of the equation in Fig. 6.9a is inserted. In the third step the first flow equation is used again. The last step finally corresponds to Eq. (6.63). Integrating the whole equation immediately leads to the Hartree self-energy.

biased in the sense that it overestimates ordering tendencies. The ladder diagrams on the other hand compensate for these effects. In particular, the particle-hole ladder can be considered as the leading contribution in a $\frac{1}{N}$ expansion, with N being the dimension of the symmetry group of the spins. The generalization from SU(2) to SU(N) requires an extra quantum number attached to the pseudo fermions. In the limit $N \to \infty$ the results of an RVB mean-field theory are recovered yielding spin liquid behavior [3, 76]. Thus, the inclusion of both, the RPA graph and the crossed particle-hole graph enables us to adequately account for the competition of magnetic order and disorder in an unbiased fashion. In addition to the bare resummations in the different channels, FRG also includes mixed contributions, representing subleading orders in the respective limitation scheme.

Katanin Truncation Applied to the J_1-J_2 Model

From a numerical point of view, a significant disadvantage of the Katanin truncation is the fact that the second term of the redefined single-scale propagator, see Eq. (6.62), does not contain a δ-function that cancels the internal frequency integration. Hence, the loop integration has to be evaluated numerically by summing up small trapezoidal elements. Apart from the numerical integration, the implementation of the FRG scheme follows the explanations in Section 6.2. The two-particle vertex has three frequency arguments resulting in computation times that grow with the third power of the number of frequencies. Moreover, due to the one-dimensional internal frequency integration the computation time depends on the number of integration mesh-points. Concerning the spatial dependence the computational effort again grows with the fourth power of the length of the longest two-particle vertex. Typically, we use 64 frequencies and discard all two-particle vertices with a spatial extent longer than seven lattice spacings, i.e., we have $L = 7$ in Eq. (6.47). Exploiting lattice symmetries we end up with approximately $2.5 \cdot 10^6$ coupled differential equations. The numerically determined coupling functions and self energies are inserted into Eq. (6.54) to calculate the susceptibilities.

The numerical solution is shown in Fig. 6.11. In the course of the flow, the Néel

6 The Functional Renormalization Group: Implementation for Spin Systems

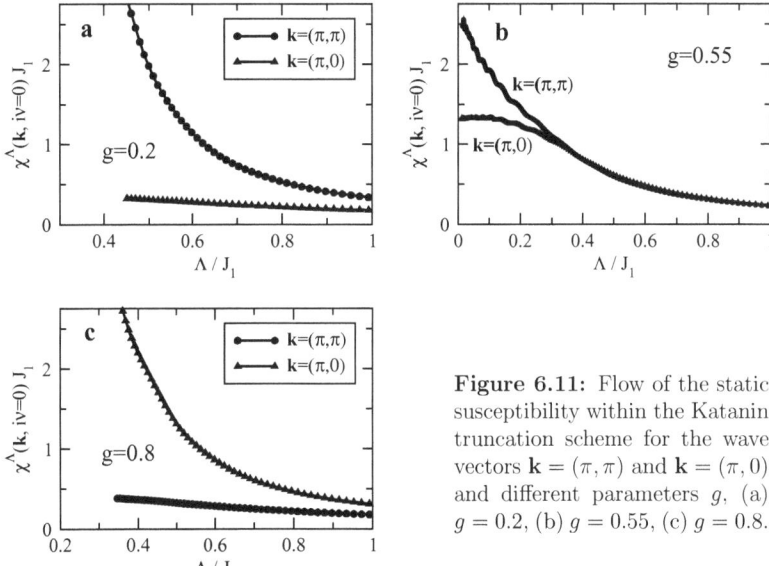

Figure 6.11: Flow of the static susceptibility within the Katanin truncation scheme for the wave vectors $\mathbf{k} = (\pi,\pi)$ and $\mathbf{k} = (\pi,0)$ and different parameters g, (a) $g = 0.2$, (b) $g = 0.55$, (c) $g = 0.8$.

susceptibility for $g = 0.2$ shows a pronounced increase while the collinear one stays very small, see Fig. 6.11a. Obviously, at that degree of frustration, the system is in the Néel phase. At $g = 0.8$ (see Fig. 6.11c) the behavior is analogous but with interchanged roles: The collinear susceptibility exhibits a strong rise while the Néel one remains small. Evidently, the collinear phase can be identified here. So far the results are similar to those obtained within the conventional truncation. A difference is seen in Fig. 6.11b at $g = 0.55$ where both susceptibilities approach finite values for $\Lambda \to 0$, demonstrating the existence of a phase with neither Néel nor collinear long-range order. Small oscillations are a consequence of the discrete frequency mesh and the linear interpolation of the vertices between adjacent mesh points. Usually, they are suppressed for a denser mesh but cannot be avoided entirely. Typically, the flow is stopped at $\Lambda \approx 0.01 J_1$ but can be easily extrapolated to $\Lambda = 0$ since no relevant processes are expected below that scale.

We emphasize that in magnetic phases we never observe real divergences of the dominant susceptibility. When Λ gets too small, the increase in the susceptibility suddenly stops and the flow exhibits an unstable and wiggly behavior. The data in Fig. 6.11a and 6.11c only represents the smooth part of the flow whereas Fig. 6.12 also shows the unstable flow. In fact, we should not expect diverging susceptibilities. Remind that the generic outcome of our FRG scheme are the spin-spin correlations in real space. A single correlator of two spins with a specified distance should not diverge on an approach to a magnetic phase. Divergences in ordered regimes occur because the correlations do

6 The Functional Renormalization Group: Implementation for Spin Systems

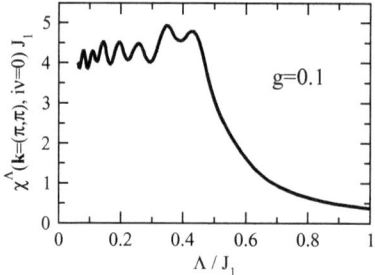

Figure 6.12: Flow of the static Néel susceptibility within the Katanin truncation scheme for $g = 0.1$ showing also the unphysical part of the flow.

not decay to zero in the limit of long distances. Hence, only for infinitely large systems the Fourier transform of the correlation function diverges at the corresponding ordering vector. On the other hand, we keep only a limited range of spin-spin correlations and consequently the Fourier transform includes a finite site summation that cannot diverge. Nevertheless in magnetic phases the flow enters a regime where typical ordering length-scales become larger than the extent of the longest correlations. As illustrated in Fig. 6.12, below that Λ-scale the flow is unphysical (in Fig. 6.12 below $\Lambda \approx 0.45$): Here the susceptibilities depend very sensitively on the discretization of the frequencies, leading to large oscillations. A different mesh changes the flow behavior considerably. Interestingly the characteristic step at $\Lambda \approx 0.45$ is smoothened if one omits small frequencies in the mesh. In that case the Goldstone mode is insufficiently resolved such that critical behavior is suppressed. Furthermore finite size effects are not negligible. Considering larger system sizes one can follow the flow to larger susceptibilities and finds a steeper increase before the breakdown. These observations demonstrate that the flow is badly converged below the critical Λ-scale which is consistent with an ordering instability. However, in the thermodynamic limit and with a sufficient number of discrete frequencies we expect a smooth diverging solution.

We also note that within the conventional truncation of Section 6.4 we obtain real divergences of the susceptibilities. Even though this is the behavior one would naively expect on an approach to a magnetic instability, it is an unphysical artefact of the improper approximation because there the spin-spin correlations diverge individually. The oscillations in the flow which are a consequence of the frequency discretization and the interpolation between the mesh points may be considered as a measure for the stability of the solution and hence also for the size of the magnetic fluctuations. In fact, far away from ordered phases, i.e., at $g \approx 0.55$, the oscillations are the smallest.

In order to further investigate the properties of the paramagnetic phase, we have calculated the susceptibilities at additional parameter values. The results in the physical limit $\Lambda = 0$ are shown in Fig. 6.13. Deep inside the paramagnetic phase our results are well-converged. With increasing g we observe a decreasing Néel susceptibility and an increasing collinear susceptibility. The point where Néel-like fluctuations loose out compared to collinear fluctuations lies at $g \approx 0.6$ in correspondence with the results of the static FRG approach in Section 6.3, i.e., clearly higher than the classical value

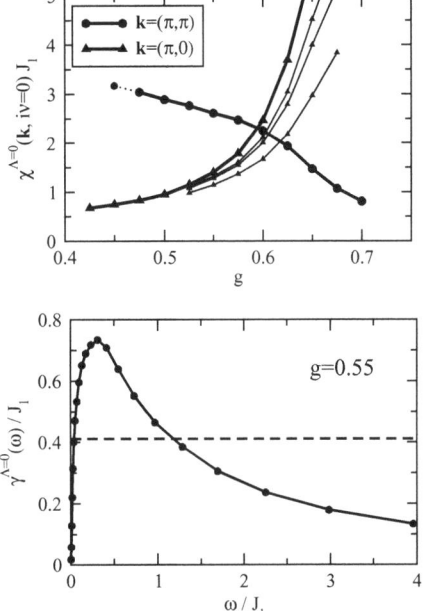

Figure 6.13: Static susceptibility for the paramagnetic phase in the physical limit $\Lambda = 0$. The thick lines are obtained by a finite size scaling. The thin lines are the results for maximal bond lengths with $L = 7$, $L = 5$ and $L = 3$ (see Eq. (6.47)), from top to bottom.

Figure 6.14: Frequency dependent damping $\gamma^{\Lambda=0}(\omega)$ obtained at the end of the FRG-flow compared to the constant damping with $\tilde{\gamma} = 0.36$ used in Chapter 5 (dashed line). Depicted is the case $g = 0.55$.

$g = 0.5$. Near the phase boundary to collinear order, which turns out to be in the range $g_{c2} \approx 0.66\ldots0.68$, the collinear susceptibility shows a pronounced increase until at $g \gtrsim g_{c2}$ a characteristic step as in Fig. 6.12 emerges. In that parameter regime we need large system sizes, in order to obtain well-converged results. A finite size scaling (see the thin lines in Fig. 6.13) considerably enhances the collinear susceptibility such that a beginning divergence is visible. The situation is very different near the phase boundary to Néel order. A divergence of the Néel susceptibility is not seen and finite-size effects play a minor role. Approaching the paramagnetic phase from below the step-like instability breakdown smooths continuously without any significant change in the susceptibility at $\Lambda = 0$. Therefore, it seems to be difficult to access the critical region and to obtain reliable data, see the dotted part of the (π, π)-line in Fig. 6.13. An estimation of the phase boundary leads to the parameter region $g_{c1} \approx 0.4\ldots0.45$. Both transition points are in good agreement with previous studies [43, 46, 62, 109, 123, 131]. So far we notice that the behavior of the system near the two phase boundaries is very different. To draw a conclusion concerning the order of the transitions, however, the present investigation is not sufficient.

Even though the damping γ^Λ is no physical observable, it is interesting to study this quantity, especially in comparison with the phenomenological spectral broadening from Chapter 5. The damping which is related to the self energy via Eq. (6.35) is directly generated by the first flow equation. Fig. 6.14 shows the FRG solution $\gamma^{\Lambda=0}(\omega)$ together with the frequency-independent damping $\gamma = 0.36 J_1 \sqrt{1+g^2}$ used previously. The FRG result falls off with increasing frequency (typically as $\frac{1}{\omega}$ for large enough frequencies), however, in the non-magnetic phase at $g = 0.55$ we find that both quantities compare quite well at relevant energy scales $\omega \sim J$.

To conclude, in this section we demonstrated that the inclusion of certain higher-order terms in the RG equation for Γ as depicted in Fig. 6.8 turns out to be essential for the competition between order and disorder. The damping has indeed just the right size to reproduce the phase diagram found in previous studies. We repeat the mechanism that leads to the correct results: Unlike the conventional truncation, the Katanin scheme assures a sufficient feedback of the self energy into the two-particle vertex. This is achieved by the extra term in the single-scale propagator, see Eq. (6.62), which guarantees that the dressed RPA is generated during the flow. According to the discussion in Chapter 5, in that case, the damping controls the disorder fluctuations of the system. However, the RPA considered separately is finite only in the spin-interaction sector but not in the density-interaction sector. Hence, it does not generate a finite self energy if coupled back into the Hartree term of the first flow equation. To say this in other words, the RPA + Hartree approximation is the leading term in a $\frac{1}{S}$ expansion and therefore overestimates order tendencies. The other graphs such as the Fock term of the first flow equation as well as the crossed particle-hole and particle-particle terms of the second flow equation compensate for this bias since they produce a finite damping. Especially, the crossed particle-hole ladder which is the leading order in a $\frac{1}{N}$ expansion can be viewed as the counterpart of the RPA term because it favors spin-liquid behavior. Concerning both contributions, the full, dressed diagram series are included which assures that order and disorder are treated on equal footing. This generates the correct pseudo-fermion damping.

6.6 Dimer and Plaquette Order

Before we turn to the discussion of further spin systems, in this section we present a technique that allows us to characterize the nature of paramagnetic phases. Even if a system is magnetically disordered there might still be some kind of hidden long-range order in the form of a spontaneous dimerization with a certain singlet-bond pattern. Concerning the J_1-J_2 model, several states are currently debated: Firstly, a VBS with a columnar dimer arrangement which breaks translation invariance along one lattice direction as well as rotation symmetry and secondly a plaquette VBS for which the dimerization takes place on units of 2×2 plaquettes. The latter breaks translation symmetry in both directions while the rotation symmetry is intact. However, as a third alternative, the system might also be in a spin-liquid ground state that does not exhibit any kind of broken symmetries. Until now, there is remarkable disagreement in the literature concerning these three possibilities.

In order to probe the paramagnetic phase with respect to these states we add a small dimer-field perturbation to the Hamiltonian and investigate the response [28, 30, 43,

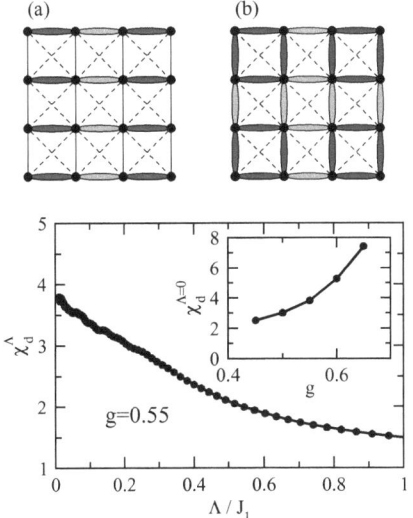

Figure 6.15: Patterns for (a) columnar dimerization and (b) plaquette dimerization. Dark gray bonds relate to strengthened and light gray bonds to weakened interactions in the Hamiltonian $H + F_\mathrm{d}$ or $H + F_\mathrm{p}$

Figure 6.16: Main figure: Flowing dimer correlation χ_d^Λ for $g = 0.55$. Inset: Dimer correlation $\chi_\mathrm{d}^{\Lambda=0}$ in the limit $\Lambda = 0$ versus g.

62, 125, 131, 132]. In the context of FRG this concept has already been applied in Refs. [51, 107]. The fields can be chosen as

$$F_\mathrm{d} = \delta \sum_{i,j} (-1)^i \mathbf{S}_{i,j} \mathbf{S}_{i+1,j} ,$$
$$F_\mathrm{p} = \delta \sum_{i,j} \left[(-1)^i \mathbf{S}_{i,j} \mathbf{S}_{i+1,j} + (-1)^j \mathbf{S}_{i,j} \mathbf{S}_{i,j+1} \right] , \quad (6.68)$$

for the columnar dimer and plaquette order, respectively. Here i, j are components of the position vector and δ is an energy much smaller than J_1 and J_2. Note that the expectation values $\langle F_\mathrm{d} \rangle$ and $\langle F_\mathrm{p} \rangle$ are the order parameters of these states. F_d (see Refs. [28, 30, 43, 62, 125, 131, 132]) and F_p (see Refs. [43, 62]) break the above-mentioned lattice symmetries and generate the two dimerization patterns shown in Fig. 6.15. Possible instabilities should be visible as divergences in the corresponding equal-time correlation functions $\chi_\mathrm{d/p} = \frac{d\langle F_\mathrm{d/p}\rangle}{d\delta}|_{\delta=0}$.

A perturbation in the form of two-body spin fields as shown in Eq. (6.68) amounts to a slight strengthening or weakening of the interactions in the initial conditions, according to the dimerization patterns. In the course of the flow we calculate the correlations of strengthened and weakened bonds and its relative difference. We define equal time dimer and plaquette correlation functions by

$$\chi_\mathrm{d/p}^\Lambda = \frac{J_1}{\delta} \frac{\left| \left(\langle\langle \mathbf{S}_{i,j}, \mathbf{S}_{i+1,j} \rangle\rangle_\mathrm{d/p}^\Lambda - \langle\langle \mathbf{S}_{i+1,j}, \mathbf{S}_{i+2,j} \rangle\rangle_\mathrm{d/p}^\Lambda \right) \right|}{\left(\langle\langle \mathbf{S}_{i,j}, \mathbf{S}_{i+1,j} \rangle\rangle_\mathrm{d/p}^\Lambda + \langle\langle \mathbf{S}_{i+1,j}, \mathbf{S}_{i+2,j} \rangle\rangle_\mathrm{d/p}^\Lambda \right)} . \quad (6.69)$$

Here the index d/p indicates that the correlator $\langle\langle\ldots\rangle\rangle$ is calculated with the Hamiltonian $H + F_{\rm d/p}$. The factor δ in the denominator eliminates the dependence on the strength of the perturbation such that the flow of $\chi_{\rm d/p}^\Lambda$ starts with the initial value $\chi_{\rm d/p}^{\Lambda=\infty} = 1$. An increase (decrease) during the flow shows that the system supports (rejects) the perturbation. Note that we again apply Katanin's truncation scheme.

The columnar dimer correlation $\chi_{\rm d}^\Lambda$ is plotted in Fig. 6.16. It is seen that this quantity increases considerably during the flow. At $g = 0.55$ the perturbation is enhanced by a factor of ≈ 3.8 in the limit $\Lambda \to 0$ but a divergence does not occur. Even larger responses are obtained with increasing g, see the inset of Fig. 6.16. Remarkably we obtain plaquette correlations $\chi_{\rm p}^\Lambda$ with the same strength. Hence, the FRG scheme is apparently not able to distinguish between dimer and plaquette correlations. We do not exclude instabilities for this types of order: In fact, an unambiguous detection of a valence-bond dimerization requires the explicit inclusion of dimer susceptibilities in the form of the four-particle vertex. This is, however, beyond the scope of the present FRG formulation. On the basis of our approach we can at least distinguish between regimes of different dimer-fluctuation strengths and test different patterns against each other.

In conclusion we presented a method that allows to estimate dimer fluctuations in paramagnetic phases. Applied to the J_1-J_2 model we find enhanced columnar dimer and plaquette responses of equal strength. Especially near the transition to the collinear ordered phase a dimerization is favored. However, a clear detection of such instabilities requires further diagrammatic contributions. In Section 7.5 we revisit this scheme in the context of the Heisenberg model on the honeycomb lattice.

7 Application to Further Models

Our findings in the previous chapter led to the conclusion that FRG with pseudo fermions in conjunction with the Katanin truncation scheme is capable of reproducing the correct phase diagram of the J_1-J_2 Heisenberg model. We demonstrated that the competition between order and disorder is properly described, yielding reasonable values for the transition points. In the present chapter we will answer the question if our method is equally powerful when applied to other spin models with different arrangements of the interactions or different underlying lattices. In principle, a generalization of such kind may be implemented straightforwardly since a modification of the interactions only requires an adjustment of the initial conditions. The implementation of a different lattice amounts to rearranging the internal site summations. Below, we consider models of very different kind, exhibiting rich phase diagrams, but all these models have in common that order and disorder tendencies compete against each other such that paramagnetic phases are obtained. A first generalization of the J_1-J_2 model, known as the J_1-J_2-J_3 model, is investigated in Section 7.1 where we add a third-nearest neighbor interaction. In Section 7.2, we consider the Heisenberg model on a checkerboard lattice which is similar to the J_1-J_2 model but with an arrangement of interactions that enlarges the unit cell. After these discussions of square-lattice models it appears to be natural to address systems on a triangular lattice. Firstly, in Section 7.3 we study the Heisenberg model on a bare triangular lattice, also allowing for anisotropic interactions in real space. We continue examining non-Bravais like lattices in the following two sections: The Kagome lattice, see Section 7.4, is constructed by omitting certain sites of the triangular lattice which enhances frustration effects. Thereafter, the honeycomb lattice with interactions up to the third-nearest neighbors is considered in Section 7.5. Finally, Section 7.6 investigates a generalization of the Heisenberg model on the honeycomb lattice which also includes ferromagnetic interactions and anisotropies in spin space. This model is known as the Kitaev-Heisenberg model. In particular, the systems described in Sections 7.3, 7.4 and 7.6 are interesting also from an experimentalists point of view as there exist materials or at least promising candidates for such model Hamiltonians.

7.1 The J_1-J_2-J_3 Heisenberg Model

One approach to shed additional light on the phase diagram of the J_1-J_2 model is to embed its analysis into a larger parameter space. In this context the J_1-J_2-J_3 model [99]

$$H = J_1 \sum_{\langle i,j \rangle} \mathbf{S}_i \cdot \mathbf{S}_j + J_2 \sum_{\langle\langle i,j \rangle\rangle} \mathbf{S}_i \cdot \mathbf{S}_j + J_3 \sum_{\langle\langle\langle i,j \rangle\rangle\rangle} \mathbf{S}_i \cdot \mathbf{S}_j \qquad (7.1)$$

7 Application to Further Models

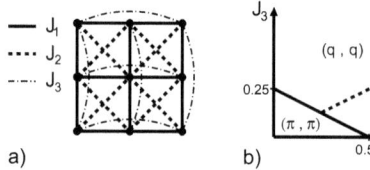

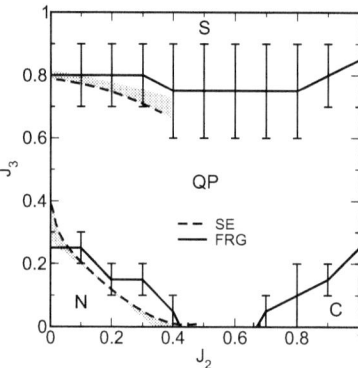

Figure 7.1: a) J_1-J_2-J_3 model. Solid dots refer to the lattice sites. b) The classical phase diagram of the J_1-J_2-J_3 model.

Figure 7.2: Quantum phase diagram of the J_1-J_2-J_3 model. Solid: Onset of magnetic flow behavior from FRG. Error bars of size 0.1 (larger than 0.1) are due to the finite $J_{2,3}$ mesh (uncertainties in the flow), see text. Dashed: Triplet-gap-closure from 5th order series expansion [11]. The shaded regions refer to the difference between bare series and DlogPadé, for details, see [11]. "N", "C" and "S" denote the Néel, collinear and spiral state. "QP" refers to a quantum paramagnet.

has become of renewed interest recently. The spin operators are located on the sites of the planar square lattice shown in Fig. 7.1, and $J_{1,2,3} \geq 0$ are antiferromagnetic exchange couplings ranging from first, i.e. $\langle i,j \rangle$, up to third-nearest neighbors, i.e. $\langle\langle\langle i,j \rangle\rangle\rangle$. The J_1-J_2 model is obviously recovered in the case $J_3 = 0$. In the following we set $J_1 = 1$.

Classically, the J_1-J_2-J_3 model allows for four ordered phases [33, 47], comprising a Néel, a collinear, and two types of spiral states which are depicted in Fig. 7.1. Except for the transition from the diagonal (q,q)-spiral to the (π,q)-spiral state, which is first order, all remaining transitions are continuous. Early analysis of quantum fluctuations [47] found the Néel phase to be stabilized by $J_3 > 0$, with the end-point of the classical critical line $J_3 = 1/4 - J_2/2$ at $J_2 = 0$ shifted to substantially larger values of J_3. First indications of non-classical behavior for finite $J_3 > 0$ where obtained at $J_2 = 0$. A "spin-Peierls state" was found in exact diagonalization (ED) studies in the vicinity of $J_3 \sim 0.7$, between the Néel phase and the diagonal spiral [72]. Monte-Carlo and $1/N$ expansion resulted in a succession of a VBS and a Z_2 spin-liquid in this region [29]. Quantum paramagnetic behavior was also conjectured at finite J_2, J_3, along the line $J_2 = 2J_3$ using Schwinger-Bosons [33]. More recent analysis, based on ED and short-range valence-bond methods found an s-wave plaquette VBS, breaking only translational symmetry, along the line $J_2 + J_3 = 1/2$, up to $J_2 \lesssim 0.25$ [74]. This VBS's region of stability was then studied by series expansion in the (J_2, J_3) plane [11]. Results from projected entangled pair states (PEPS) at $J_2 = 0$ supported the notion of an s-wave plaquette along the J_3-axis [85]. However, the symmetry of the non-magnetic state remains under scrutiny,

7 Application to Further Models

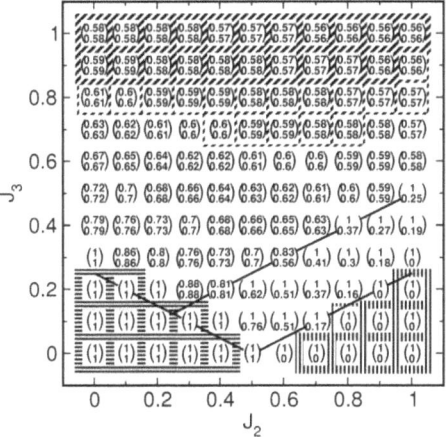

Figure 7.3: Brackets: wave vector (k_x, k_y) in units of π at the maximum of the static susceptibility from FRG. Solid lines indicate the classical critical lines. Horizontally, vertically and diagonally striped frames around the brackets correspond to Néel, collinear, and spiral states, respectively. Thin frames mark regions of uncertain flow behavior.

since a truncated quantum-dimer model [92] indicates that the potential plaquette VBS has a subleading columnar dimer admixture in the vicinity of $J_2 \approx J_3 \approx 0.25$, similar to ED studies [119]. This implies broken translation and rotation symmetry. On the other hand, for $J_2 \gtrsim 0.5$, ED shows strong columnar dimer correlations [119]. Finally, the order of the transitions from the quantum paramagnetic into the semiclassical phases, and in particular to the diagonal spiral, remain an open issue.

Using FRG in conjunction with the Katanin truncation in the way described in Sections 6.2 and 6.5, we intend to further clarify the extent of the paramagnetic region. The phase diagram has been calculated in the J_2-J_3-plane with steps of 0.1 for $0 \leq J_{2,3} \leq 1$. Due to the two-dimensional parameter space a large computational effort is required to locate the phase boundaries with sufficient accuracy. For the present calculation we have used 46 frequency points. The spatial dependence of the spin-correlation function was kept up to lattice vectors $\mathbf{R}$ satisfying $\text{Max}(|R_x|, |R_y|) \leq 5$ and the correlations were put to zero beyond that range. This proved to be sufficient for a first exploration of the phase diagram. As described in Section 6.5, in magnetic phases we see a pronounced susceptibility peak in momentum space that rapidly grows during the Λ-flow. At a certain Λ the onset of spontaneous long-range order is signalled by a sudden stop of the smooth increase and the onset of an unstable, oscillating flow behavior. On the other hand, in non-magnetic phases a smooth flow and broad susceptibility peaks are obtained. This distinction allows us to draw the phase diagram of the model, which is shown in Fig. 7.2. Regarding the error bars in Fig. 7.2 we note that bars of size 0.1 into the J_3-direction do not reflect errors of the FRG, but are only due to the finite (J_2, J_3)-spacing and, in principle, apply also to the J_2-direction. However, especially near the phase boundary between the spiral ordered and the disordered phase, at large J_3, we encounter enhanced uncertainties. Here (J_2, J_3)-regions occur where it is not clear if the behavior of the flow should be interpreted as magnetic or non-magnetic, because the characteristic step-like

7 Application to Further Models

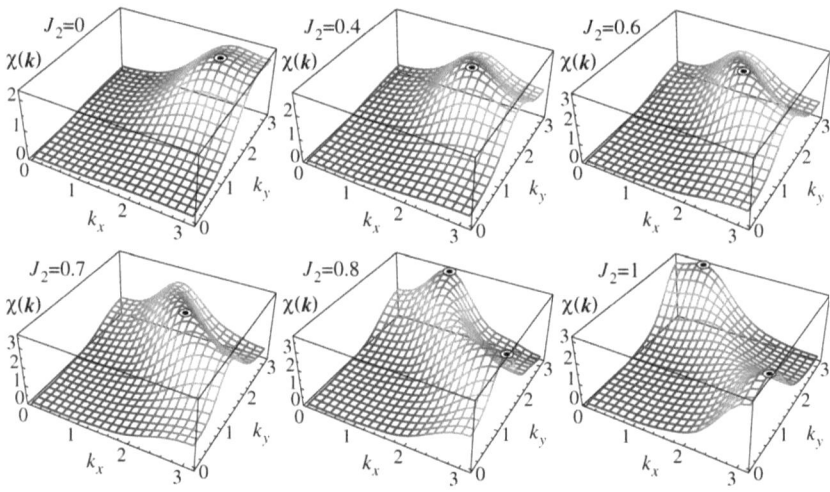

Figure 7.4: Static susceptibility for wave vectors $k_x, k_y \in [0, \pi]$ for various values J_2 and constant $J_3 = 0.4$. The black dots mark the positions of the maxima.

instability breakdown is not as pronounced as in Fig. 6.12. In Fig. 7.2 these regions lead to error bars larger than 0.1. Despite these uncertainties FRG reproduces the results of previous studies (as series expansion [11], see also Fig. 7.2) quite well.

Since FRG evaluates the static susceptibility over the complete Brillouin zone, it allows to determine the wave vector $\mathbf{k}_{\text{max}}$ of the dominant short-range magnetic correlations or the pitch vector of the magnetic order parameter. These wave vectors are depicted in Fig. 7.3 together with the quantum phases discussed already in Fig. 7.2. Both, in the ordered as well as in the paramagnetic phase we find the wave vectors at the maximum of the susceptibility to agree approximately with those obtained for the purely classical model in Fig. 7.1. This is particularly interesting with respect to the (π, q)-spiral state, which seems to exist only in the form of short range correlations in Fig. 7.3.

In order to illustrate how the dominant fluctuations in the disordered phase change with varying couplings, in Fig. 7.4 we show results for the static susceptibility as a function of the k-vector in the Brillouin zone with $k_x, k_y \in [0, \pi]$ at fixed $J_3 = 0.4$ for various values of J_2. For $J_2 = 0$ we see a broadened peak at a (q, q)-position which has already moved away from the Néel-point $\mathbf{k} = (\pi, \pi)$. This peak further moves along the Brillouin-zone diagonal for increasing J_2. For $J_2 \gtrsim 0.6$ it is seen that the peak smoothly deforms into an arc and that the weight at the Brillouin-zone boundary increases. Between $J_2 = 0.7$ and $J_2 = 0.8$, close to the classical first-order transition, the ridge has constant weight and the maximum jumps to a (π, q)-position to then further evolve towards the collinear points $\mathbf{k} = (0, \pi)$, $\mathbf{k} = (\pi, 0)$ and to acquire more

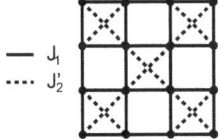

Figure 7.5: The checkerboard lattice. Dots indicate the spins of the Heisenberg model.

prominence. Therefore, remnants of the classical correlations survive into the non-magnetic regime.

In conclusion, we find a phase diagram of the J_1-J_2-J_3 model with a large paramagnetic region, which is in good agreement with previous studies. This demonstrates that the FRG is capable to describe the interplay of different phases even if more complicated types of order such as spiral order are involved. Moreover, it is shown that the k-space resolved susceptibility allows to identify the dominant order fluctuations in the paramagnetic phase. These fluctuations are found to be remnants of the classical order.

7.2 The Heisenberg Model on a Checkerboard Lattice

Instead of adding further exchange couplings as done in the previous section for the J_1-J_2-J_3 model we now consider a different modification of the J_1-J_2 model, referred to as the checkerboard lattice. Compared to the J_1-J_2 model, in one half of the square plaquettes the diagonal couplings J_2 are omitted such that empty plaquettes and plaquettes with diagonal interactions form a checkerboard pattern, see Fig 7.5. In the literature this system is also called the planar pyrochlore or the crossed chain model. The Heisenberg Hamiltonian of the system is given by

$$H = J_1 \sum_{\langle i,j \rangle} \mathbf{S}_i \cdot \mathbf{S}_j + J_2' \sum_{\langle\langle i,j \rangle\rangle'} \mathbf{S}_i \cdot \mathbf{S}_j \,, \tag{7.2}$$

where the sum $\sum_{\langle i,j \rangle}$ runs over the usual nearest-neighbor pairs on a square lattice while $\langle\langle i,j \rangle\rangle'$ denotes next-nearest pairs on one half of the square plaquettes as indicated in Fig. 7.5. The dimensionless parameter $g' = \frac{J_2'}{J_1}$ spans the phase diagram of the system.

One motivation for studying this model comes from materials such as $Na_2Ti_2Sb_2O$ and $Na_2Ti_2As_2O$ for which the Hamiltonian (7.2) is believed to give an appropriate description [122]. The oxygen atoms form a square lattice and the titanium atoms sit at the bond centers such that they are also arranged in a square lattice. The spin-half magnetic moments are provided by the titanium atoms. A direct overlap between the titanium orbitals yields nearest neighbor couplings J_1 while an interaction mediated by the oxygen orbitals results in the couplings J_2'.

The theoretical interest in the model stems from the fact that the frustrating arrangement of the interactions leads to a melting of magnetic order. It is well known that the system features a Néel phase for small g', similar to the J_1-J_2 model. However, due to the smaller number of diagonal couplings, a larger J_2' is needed to melt the magnetic order. Both, spin-wave theory [26, 122] and exact diagonalization [117] consistently find that the Néel order survives up to a critical coupling $g'_c \approx 0.75$. There is general agreement that beyond that value the system is in a non-magnetic valence bond solid phase with a large spin gap to triplet excitations of the order of the exchange couplings [18, 21, 26, 48, 89, 117]. It is reported that this valence-bond ground state may be well approximated by a product of disconnected singlet plaquettes on one half of the non-crossed unit squares [18, 21, 26, 48, 117]. The spin gap itself is filled with a large number of singlet states [21, 48, 89], which, however, do not form a continuum above the ground state. There is indication that also the singlet excitations are gapped [48]. According to exact diagonalization the valence-bond solid phase includes the isotropic point $g' = 1$ [117]. This point is of special importance as explained below. As one further increases g', the diagonal J'_2-bonds effectively form crossed spin chains which are weakly coupled by the frustrating interaction J_1. The limit $g' \to \infty$ where the chains are completely decoupled is well understood: Each spin chain is in an ungapped, non-dimerized spin-liquid state with deconfined spinons as the elementary excitations. However, for a small but finite interchain coupling J_1 the situation is less clear. Early random phase approximation (RPA) [130] and exact diagonalization studies [117] conclude that the 1D spin-liquid behavior remains intact when a small J_1 is switched on. They consider the checkerboard lattice as a good candidate for a 2D system which retains 1D Luttinger physics and call the corresponding phase the "sliding Luttinger liquid phase". Exact diagonalization finds a rather large extent of this state, i.e., down to $g' \approx 1.25$ where it contacts the aforementioned valence-bond solid [117]. In contrast, quite recent studies based on the renormalization group method and one-dimensional bosonization come to a different conclusion. They find a dimerized state in the large g'-regime with a crosswise arrangement of singlets in the unit squares with diagonal interactions [129].

The isotropic point $g' = 1$ is of special interest since frustration effects are expected to be exceptionally strong there. At this point the system can be considered as the 2D version of the pyrochlore lattice which is built up of corner sharing tetrahedra forming a 3D lattice altogether. Despite the difficulties in the numerical treatment of 3D spin models it is a good candidate for a quantum spin liquid. The checkerboard model at $g' = 1$ has often been perceived as a simplified system for the pyrochlore lattice since its main features are retained: Firstly, the unit squares with diagonal J'_2-bonds have the connectivity of tetrahedra and secondly, these tetrahedra are only linked at their corners such that the local environment of each site is the same as in the pyrochlore lattice. We emphasize that systems with triangular building blocks (i.e., triangles or tetrahedra) coupled in a corner sharing fashion generally exhibit large frustration effects. Note that the Kagome lattice that will be studied in Section 7.4 also falls into this class of systems.

We mention that the corner sharing property has an interesting implication on the corresponding classical large-spin model. Generally, on these lattices the Hamiltonian can be rewritten as the sum of the squares of the total spin of corner sharing units α,

$$H = J \sum_{\langle i,j \rangle} \mathbf{S}_i \cdot \mathbf{S}_j = \frac{J}{2} \sum_\alpha \mathbf{S}_\alpha^2 + \text{const.}, \qquad (7.3)$$

where $\mathbf{S}_\alpha$ is the total spin in the unit α. Thus, whenever $\mathbf{S}_\alpha = 0$ for each α, a classical ground state is obtained, resulting in a macroscopic classical degeneracy. However,

7 Application to Further Models

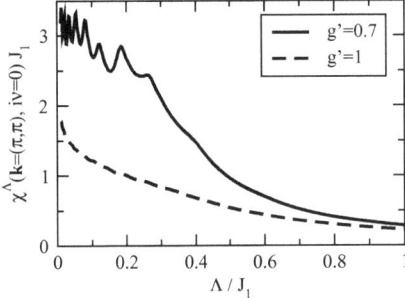

Figure 7.6: Flow of the static Néel susceptibility in the checkerboard model for $g' = 0.7$ and for the isotropic case $g' = 1$.

as pointed out in Ref. [48], this degeneracy is lifted in the quantum case and is not a sufficient condition for a continuum of singlet states in the corresponding quantum model.

In our FRG study presented below we concentrate on the regime where g' is not too large, i.e., $0 \leq g' \leq 2$. The weakly coupled chain limit $g' \gg 1$ is hardly accessible within our approach as will be exemplified in Chapter 10. Concerning the method, we again follow the route described in Sections 6.2 and 6.5 with the only exception that we have to account for the polyatomic unit cell of the lattice (the checkerboard lattice has a two-atomic unit cell). In this case the two-particle vertex is not uniquely determined by the distance vector $\mathbf{R}_{i_1} - \mathbf{R}_{i_2}$ and moreover it is no longer an even function in the transfer frequencies s and u, see the considerations in appendix C. Instead the two-particle vertex is invariant under the combined transformations $s \to -s$, $i_1 \leftrightarrow i_2$ and $u \to -u$, $i_1 \leftrightarrow i_2$ (the invariances under $t \to -t$ and $s \leftrightarrow u$ remain valid). Since our FRG algorithm uses only positive frequencies s, t, u, the transformation of a two-particle vertex into this sector possibly requires an exchange of the sites, i.e., $i_1 \leftrightarrow i_2$. However, this operation can be easily implemented in our computer code.

Furthermore, the two-atomic unit cell has to be accounted for when we calculate the $\mathbf{k}$-space resolved susceptibility $\chi^\Lambda(\mathbf{k})$. Inserting Eq. (6.56) into the definition of $\chi^\Lambda(\mathbf{k})$, Eq. (6.55), we now obtain

$$\chi^\Lambda(\mathbf{k}, i\nu) = \frac{1}{2} \sum_{i \in \alpha} \sum_j e^{i\mathbf{k}(\mathbf{R}_i - \mathbf{R}_j)} \chi_{ij}^\Lambda(i\nu), \qquad (7.4)$$

where the sum $\sum_{i \in \alpha}$ runs over all sites of an arbitrary unit cell α. The prefactor $\frac{1}{2}$ is the inverse of the number of sites in a unit cell. Note that $\chi^\Lambda(\mathbf{k})$ is not periodic with respect to the first Brillouin zone of the lattice but rather with respect to the extended Brillouin zone. The latter is constructed using the reciprocal lattice vectors of the underlying square lattice, i.e., in our convention Néel order still corresponds to a wave vector $\mathbf{k} = (\pi, \pi)$. In order to show the full information contained in $\mathbf{k}$ space we always depict the extended Brillouin zone in the following.

As we sweep through the phase diagram, beginning at small g', we first observe the expected Néel state which manifests itself in a strong rise of the static susceptibility in

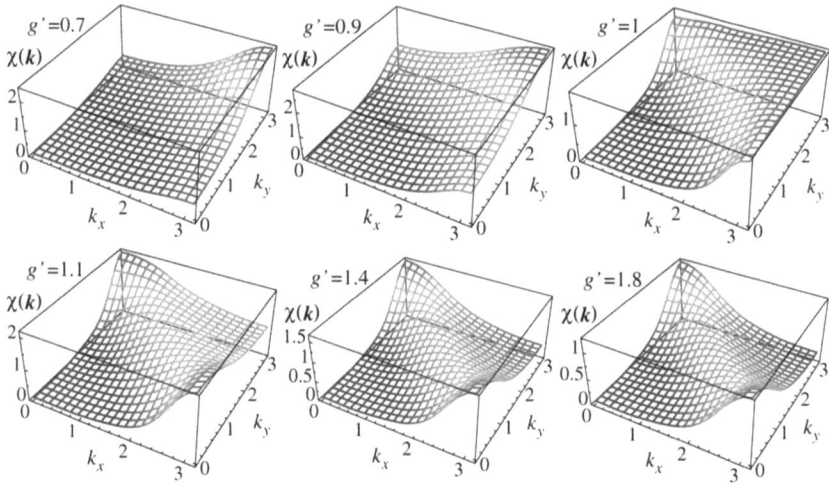

Figure 7.7: Static susceptibility for wave vectors $k_x, k_y \in [0, \pi]$ in the extended Brillouin zone upon varying g' across the transition point $g' \approx 0.75$ and the isotropic limit $g' = 1$. Note that the plot for $g' = 0.7$ shows the susceptibility just before the breakdown of the flow (i.e., at $\Lambda \approx 0.27$) while the other plots display the case $\Lambda = 0$.

the corresponding (π, π)-channel, and eventually in a sudden breakdown of the smooth flow. Even for a rather large parameter $g' = 0.7$ this behavior is still visible in the full curve of Fig. 7.6. However, because of the sizeable frustration in that regime, the breakdown and the step-like behavior is not as pronounced as for smaller g'. Also the Λ-scale of the breakdown is reduced (for $g' = 0.7$ this scale is found as $\Lambda \approx 0.27$). In a k-space plot of the first quadrant of the extended Brillouin zone the frustration effects manifest itself in a broadened peak at $\mathbf{k} = (\pi, \pi)$, see the upper left plot in Fig. 7.7 which displays the susceptibility at $\Lambda \approx 0.27$, i.e., just before the breakdown of the flow. (We emphasize that even in an unfrustrated system it cannot be expected that the susceptibility possesses sharp δ-peaks as the notion of magnetic long-range order would imply. On the one hand, this is certainly a consequence of the finite size of the correlation area. On the other hand, also the discrete frequency meshes smear out the response peaks.)

In fact, between $g' = 0.7$ and $g' = 0.8$ the breakdown of the flow smoothes such that we can pinpoint the phase transition to $g' \approx 0.75$, which agrees well with spin-wave theory [26, 122] and exact diagonalization [117]. Note that a phase transition manifests itself in the flow behavior rather than in the peak structure in k-space. Inside the paramagnetic phase ($g' \gtrsim 0.75$) the peak at $\mathbf{k} = (\pi, \pi)$ further broadens until in the isotropic limit $g' = 1$ the susceptibility has (almost) constant weight at the boundaries of the Brillouin zone, see Fig. 7.7 (all plots in Fig. 7.7 except the one for $g' = 0.7$ show the susceptibility at $\Lambda = 0$). At this point one cannot even identify residual magnetic fluctuations because any peak structure has completely disappeared, i.e., Néel- and collinear fluctuations are almost equally strong. In other words, the spin-spin correlations are of very short-range type as one would expect for a system with a large spin gap. Correspondingly, the susceptibility shows a very smooth flow without any sign of a magnetic instability, see Fig. 7.6. The isotropic limit $g' = 1$ is in fact a special point in the phase diagram as seen in Fig. 7.7: For $g' > 1$ the susceptibility at $\mathbf{k} = (\pi, \pi)$ rapidly decreases, leaving behind residual peaks in the collinear susceptibility-channel. However, the course of the susceptibility flow in those parameter regimes gives no indication for an instability of this kind.

We did not yet calculate dimer responses for the checkerboard lattice which we defer to future works. On the basis of our spin-susceptibility data we can at least say that there is no qualitative change as we cross the value $g' \approx 1.25$ where the plaquette order is expected to break down [117].

In conclusion our FRG calculation reproduces the correct transition point between the Néel phase and the paramagnetic phase. Above this point we obtain reasonable fluctuation profiles that do not favor any specific $\mathbf{k}$-vector at the maximally frustrated point $g' = 1$ but show tendencies for collinear fluctuations when $g' > 1$.

7.3 The Anisotropic Triangular Heisenberg Model

The antiferromagnetic Heisenberg model on a triangular lattice [97] represents a prototype of a geometrically frustrated quantum many-body system. Due to the geometry of the lattice, frustration is generated even if no next-nearest neighbor interactions are present. In this section we consider the Heisenberg model on a triangular lattice with spatially anisotropic interactions (ATLHM), however, special emphasis will be laid on the isotropic case. The ATLHM attracted considerable attention in recent years as an experimentally accessible testing ground for quantum-magnetism disorder phenomena. The Hamiltonian is given as

$$H_{\text{ATLHM}} = J \sum_{\langle i,j \rangle_\vee} \mathbf{S}_i \cdot \mathbf{S}_j + J' \sum_{\langle i,j \rangle_-} \mathbf{S}_i \cdot \mathbf{S}_j, \qquad (7.5)$$

where the coupling J' applies to the bonds along (horizontal) one-dimensional chains and J is the coupling between them, forming a triangular lattice altogether (Fig. 7.8a). We define $\xi = J'/J$ as a parameter to interpolate between the effective square-lattice limit $\xi = 0$ and the disordered isolated chain limit $\xi \to \infty$.

Experiments on Cs_2CuCl_4 ($\xi \sim 2.94$) provide an excellent testing ground of discussing various features of spin-liquid behavior [39]. (Influences such as Dzyaloshinskij-Moriya interactions complicate the experimental picture.) The formation of a magnetically ordered state for smaller ξ opposed to disorder tendencies for larger ξ can be nicely studied for the organic $\kappa - (\text{BEDT} - \text{TTF})_2\text{X}$ family. While $X = Cu_2[N(CN)_2]Cl$ shows

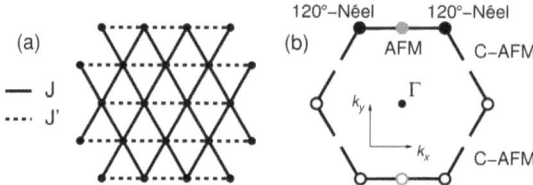

Figure 7.8: a) The triangular lattice structure. The dashed, horizontal bonds correspond to couplings J' the others to couplings J. b) Schematic plot of the hexagonal Brillouin zone. Different magnetic order resides at different points (shown: AFM Néel order, 120° Néel order, collinear AFM order (C-AFM)). The open circles relate to the filled circles by reciprocal lattice vectors, i.e., AFM order corresponds to two points in the Brillouin zone, collinear AFM to four points, and 120° Néel order to six points.

an AFM transition of $T_N = 27$K with estimated $\xi = 0.55$, the X = $Cu_2(CN)_3$ compound, estimated to be nearby the symmetric triangular regime $\xi = 1.15$, does not show magnetic order down to very low temperatures [113]. Similar findings are obtained [63] for $EtMe_3Sb[Pd(dmit)_2]_2$, which is located around $\xi = 1.1$.

Details of the phase diagram of the ATLHM are still of current debate. It is an established fact that the system is AFM Néel-ordered for small ξ, changing to incommensurate spiral order for larger $\xi < 1$ to then smoothly evolve into the well known commensurate 120°-Néel order as the isotropic triangular limit $\xi = 1$ is reached [20, 90, 112, 141]. The latter is characterized by an angle of 120° between spins on neighboring sites while next-nearest neighbor sites have the same spin orientation. However, different methods provide differing indication about the nature and position of the transition between those phases: An intermediate disordered phase, analogous to the J_1-J_2 model, has been proposed in the literature [140], while most other works assume a direct transition, but cannot classify the transition to be of first or second order [20, 112]. Compared to the classical transition located at $\xi = 0.5$ these studies find a shift of the critical coupling up to $\xi \approx 0.59$ [112] or even up to $\xi \approx 0.8$ [20, 141]. The regime beyond $\xi = 1$ attracts considerable interest because for large ξ one encounters a system of one-dimensional spin chains which are weakly coupled in a frustrated fashion, similar to the large g'-region in the checkerboard lattice. Whether or not the magnetic order persists to the isolated-chain limit is still an open issue. Some works claim a melting of the spiral order at a critical coupling $\xi > 1$ and the onset of a non-dimerized spin-liquid extending to the $\xi \to \infty$ limit [141, 144]. It has also been proposed that this phase is subdivided into a gapped and a gapless spin liquid (with increasing ξ) [144]. However, other works indicate collinear antiferromagnetic (C-AFM) ordering even in the high anisotropy regime [20, 90, 128], which corresponds to Néel order along the strongly and one weakly coupled lattice direction and ferromagnetic order along the other weakly coupled direction. This is an interesting observation since the classical estimate would be spiral order

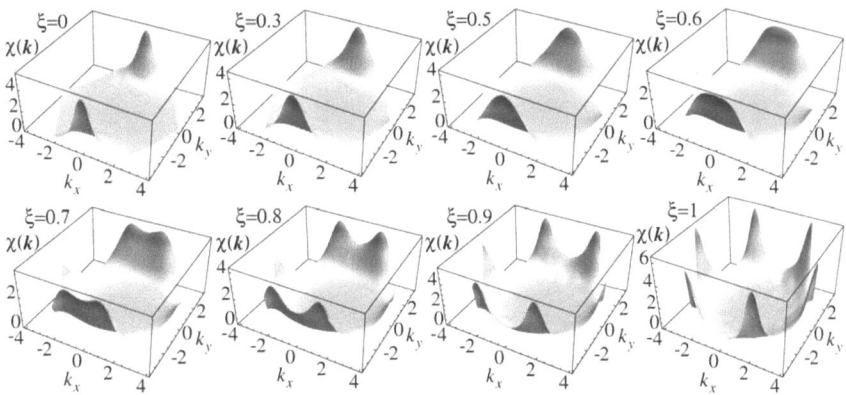

Figure 7.9: Static magnetic susceptibility resolved for the whole Brillouin zone, varying $\xi = J'/J$ from the effective square lattice $\xi = 0$ to the triangular lattice $\xi = 1$. The plots refer to the respective Λ-scales at which the magnetic instabilities occur. The peak positions for the different types of order are shown in Fig. 7.8b.

in that regime, implying that the quantum fluctuations lead to a different ordering.

Addressing this problem with FRG we sweep through the parameter space of ξ from the square lattice to the isotropic triangular lattice and compute the **k**-space resolved static magnetic susceptibility shown in Fig. 7.9. The peak positions for different types of long-range order are depicted in Fig. 7.8b. For the present calculation we used 64 discrete frequencies and included spin-spin correlations up to a distance of 7 lattice constants (note that in contrast to the quadratic correlation array as depicted in Fig. 6.4, this area is now a hexagon). Throughout this parameter regime, we observe a characteristic breakdown of the flow, indicating ordering instabilities rather than a disordered phase. One can nicely observe how the susceptibility evolves as we increase ξ. As shown in Fig. 7.9, we find that as a consequence of increasing frustration the previous Néel peak first broadens along k_x. This is followed by a splitting into two peaks which then evolve along the Brillouin-zone boundary. Simultaneously, the spectral weight at the corners of the Brillouin zone increases. Note that as a consequence of the periodicity in momentum space, the emerging peak structure at $k_y = 0$ presents the tails of the broadened Néel peak. Increasing ξ further towards the isotropic triangular point the split peaks move towards the corner position until at $\xi = 1$ the hexagonal symmetry of the susceptibility is reached. As the susceptibility evolves completely smooth through the transition, we find it to be of second order, while an extremely weak first order transition (corresponding to a slight jump of the leading susceptibility channel) cannot be excluded as a matter of principle. We identify the wave vector of the corresponding long-range ordered phases with the position of the maximal susceptibility. From this we

7 Application to Further Models

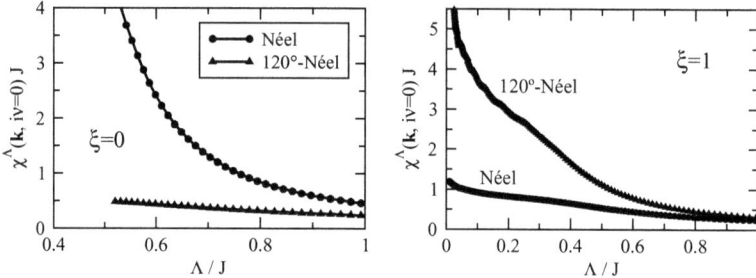

Figure 7.10: Susceptibility flow in the Néel order and 120°-Néel order channel for $\xi = 0$ (left) and $\xi = 1$ (right). For $\xi = 0$ we observe that the AFM vertices start to diverge, signalling a magnetic instability. On the other hand for $\xi = 1$ the rise of the 120°-Néel channel is seen. Compared to the flow at $\xi = 0$, the rise takes place at a much lower scale of Λ, indicating a lower ordering scale for the triangular lattice.

locate the transition point at such ξ where the peak splits and above which the order becomes incommensurate. A closer investigation using the above-mentioned system size and frequency mesh reveals a transition at $\xi \approx 0.61$. However, due to the magnetic order in that regime the flow still significantly depends on the calculation parameters. Larger sizes and more frequencies generally result in higher and sharper peaks and also in an additional shift of the transition point towards larger ξ. On the basis of data using different system sizes and frequency meshes we estimate the transition to be approximately centered between $\xi = 0.6$ and $\xi = 0.7$. In accordance with Refs. [20, 112, 141], the system influenced by quantum fluctuation favors AFM fluctuations over spiral fluctuations, since the classical transition at $\xi = 0.5$ is shifted to higher ξ.

Examples for the flow behavior at $\xi = 0$ and $\xi = 1$ can be found in Fig. 7.10. At the isotropic triangular point $\xi = 1$ where additional lattice symmetries enable us even to consider systems beyond 250 sites, finite size effects can be studied in more detail, see the discussion below.

Our results for the magnetic susceptibility at $\xi \geq 1$ are shown in Fig. 7.11. As depicted, we observe a strong drop in the magnetic susceptibility above the isotropic point, i.e., in the regime $\xi_{c2} \gtrsim 1.1$. From here, no ordering instability is found in the RG flow and the susceptibility rapidly looses the 120°-Néel order signature. While the 120°-Néel order peaks die out quickly, the leading susceptibility moves towards an incommensurate **k**-space position to then smoothly evolve into AFM stripe fluctuation signatures (at points of the Brillouin zone according to Fig. 7.8). The transition where the 120°-Néel order melts appears to be of first order according to a pronounced drop in the maximal susceptibility upon varying ξ. While we do not find a breakdown of the flow that would indicate magnetic ordering, we still obtain strong collinear AFM stripe fluctuations (in

7 Application to Further Models

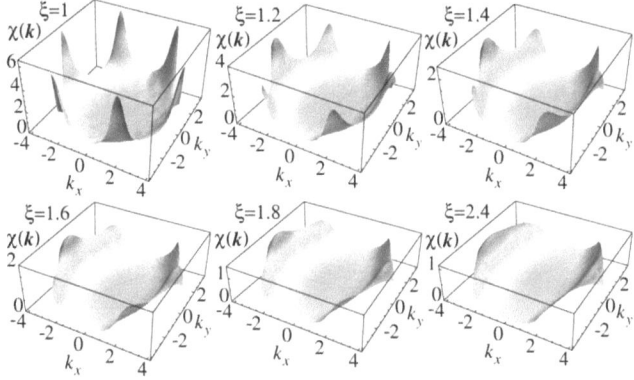

Figure 7.11: Static magnetic susceptibility, varying ξ from the isotropic triangular lattice $\xi = 1$ towards the 1D chain limit.

agreement with Ref. [128]) signalled by an unstable RG flow that develops oscillations sensitively depending on the frequency discretization. These fluctuation tendencies are also seen at higher ξ where the peak structure is still visible along the k_x-direction. However, the peaks are strongly broadened along the k_y-axis, i.e., smeared between the two C-AFM ordering-vector positions. This indicates a fast exponential decay of spin correlations between the J'-chains. Since for $\xi > 1.1$ the susceptibility evolves completely smooth, we do not find any indication for the proposals in the literature that the (supposedly) disordered regime splits into two different spin liquid phases [144].

Due to the special importance of the isotropic point $\xi = 1$ we shall discuss this case in more detail. The antiferromagnetic triangular lattice with nearest neighbor interactions is a system with a long history of controversy. For a long time, the central question was whether or not the system possesses long-range order in the form of the 120°-Néel state. The discussion was strongly influenced by an early work by Anderson [8] where he proposed the stabilization of a resonating valence-bond (RVB) ground state, i.e., a disordered spin liquid. In fact, two decades later some numerical studies such as exact diagonalization [73] seemed to support this conjecture but the issue was still far from being clarified [120]. However, over the following years more and more works claimed an ordered ground state, such that today there is general agreement that the system is in a 120°-Néel state with a magnetization of only 41% of the classical system [31]. Concerning the exact value of the magnetization there is still an ongoing debate [81].

In view of the challenge this system posed over so many years it is interesting to study how the ground state is described within FRG. As already stated above, the 120°-Néel order instability is clearly resolved during the flow. To make this statement more precise we repeated the calculation for $\xi = 1$, this time with more frequencies

7 Application to Further Models

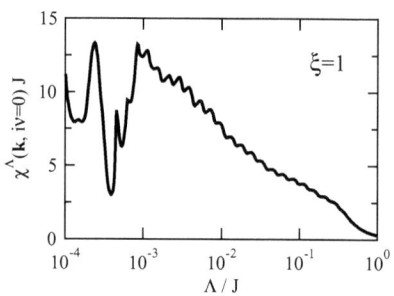
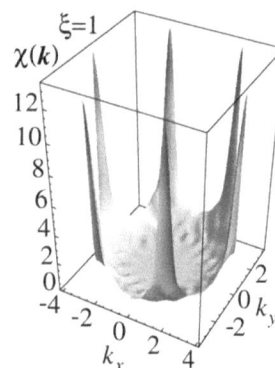

Figure 7.12: 120°-Néel susceptibility at the isotropic triangular point $\xi = 1$, logarithmically plotted as a function of Λ (left) and the **k**-space resolved susceptibility at the instability scale $\Lambda \approx 0.0008J$ (right). The calculation uses 90 frequency mesh-points and 9 lattice constants for the longest spin-spin correlations.

(90 mesh points) and a larger system size (the longest correlation extends over 9 lattice spacings, i.e., the total correlation area includes 271 sites). The computational effort at $\xi = 1$ is reduced by the fact that additional lattice symmetries may be exploited. Fig. 7.12 shows the 120°-Néel susceptibility flow for various magnitudes of Λ. It is seen that already for larger Λ magnetic fluctuations destabilize the flow and generate an oscillatory behavior. The magnetic instability eventually occurs at $\Lambda \approx 0.0008J$, i.e., much below the instability scale of the Heisenberg model on a square lattice (where the breakdown scale is $\Lambda \approx 0.45J$). This observation is consistent with the (supposedly) reduced magnetization of the system. Moreover, finite-size effects and the influence of denser frequency meshes can be nicely studied for this system. The **k**-space resolved susceptibility at the instability scale is plotted on the right side of Fig. 7.12. Compared to the corresponding plot in Fig. 7.11, it is seen that the response peaks have gained much more height, while the susceptibility profile away from the peaks remained rather unchanged (note that the relative scale of the vertical and the horizontal axes is the same in both plots). Both, the enhanced system size as well as the denser frequency mesh contribute to this effect. The former generates additional terms in the sum of the Fourier transform from real space to **k**-space and the latter leads to a better description of the Goldstone mode. On the other hand, the shift of the instability scale under these modifications is comparably small.

In conclusion we find that upon variation of the anisotropy parameter ξ the system is divided into three phases: Néel order, spiral order and a disordered phase with C-AFM fluctuations. The transition between the first two of these phases is located at $\xi \approx 0.65$ with indication to be of second order while the transition between the last two phases occurs right above the isotropic point, i.e., in the vicinity of $\xi = 1.1$ and is probably of first order.

7.4 The Heisenberg Model on a Kagome Lattice

The antiferromagnetic Heisenberg model on a Kagome lattice (see Fig. 7.13) is among the most extreme examples of frustrated spin systems realized with nearest-neighbor interactions. Similar to the checkerboard lattice, both, the frustrated triangular building blocks and the rather loose corner sharing arrangement of these units suppress the tendency towards magnetic long-range order. However, in contrast to the checkerboard lattice the Kagome antiferromagnet continues to hide the true nature of its ground state, despite the huge number of studies on that system. There are only very few statements which seem to have reached general agreement today. It is widely accepted that the ground state of the Kagome model is magnetically disordered [64, 121, 136] and that a large number of singlet excitations (possibly forming a gapless continuum above the ground state) exist at low energies [64, 118, 136]. The interpretation of this singlet continuum provides a great challenge to theorists, especially since there is no obvious type of broken symmetry. Numerical calculations suggest that triplet excitations are separated from the ground state by a small gap of about $0.05J$ [64, 136], although its existence is still being debated [118]. A lot of possible ground-state phases have been discussed in recent years but a rigorous statement cannot yet be made. Among these suggestions there are valence-bond solids with complex dimerization patterns and large unit cells of 36 sites [121]. Also non-dimerized spin liquids either with short range correlations [64] or with algebraic correlations [103] have been proposed. The latter is an interesting candidate in the case that the triplet gap vanishes. In fact, exact diagonalization [118] does not find any intrinsic energy scale and it is concluded that the system is critical or at least located near criticality. The Kagome model attracts a lot of interest also on the experimental side since materials such as herbertsmithite $ZnCu_3(OH)_6Cl_2$ provide very accurate realizations of this system. In fact, experiments on herbertsmithite show that this material is non-magnetic down to very small temperatures [56].

We performed an FRG study for the Kagome antiferromagnet and computed the static susceptibility in the extended Brillouin zone, see Fig. 7.13. Given the spin-spin correlations in real space, the susceptibility is calculated similarly to Eq. (7.4), i.e.,

$$\chi^\Lambda(\mathbf{k}, i\nu) = \frac{1}{3} \sum_{i \in \alpha} \sum_j e^{i\mathbf{k}(\mathbf{R}_i - \mathbf{R}_j)} \chi^\Lambda_{ij}(i\nu), \qquad (7.6)$$

where $\sum_{i \in \alpha}$ denotes a sum over the three sites of an arbitrary unit cell. This quantity has the periodicity of the extended Brillouin zone but not of the first Brillouin zone. Hence, susceptibilities are always presented in the former. The construction of this zone uses the lattice vectors of the bare triangular lattice such that the susceptibility profile in Fig. 7.13 can be directly compared with the $\mathbf{k}$-space resolved plots in Section 7.3. Although our FRG approach in its present version is not capable to give a detailed description of the ground state, especially if complex dimerization patterns are involved, the susceptibility

7 Application to Further Models

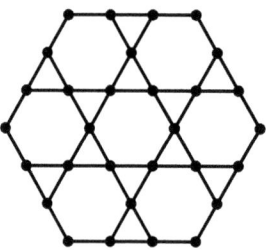
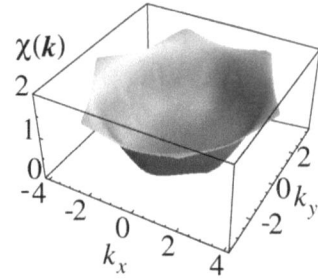

Figure 7.13: Left: The Kagome lattice. Dots indicate the locations of the spins. Right: Static susceptibility in the extended Brillouin zone using 64 frequencies and 9 lattice spacings for the longest correlations.

provides valuable information about the magnetic properties. Compared to the susceptibility of the triangular Heisenberg model in Fig. 7.12 it is remarkable that the response peaks have almost completely disappeared. The remaining small peaks have no significant dependence on the system size and the frequency discretization-mesh. Therefore we can exclude magnetic long range order. The susceptibility profile is in nice correspondence with exact diagonalization [71] with the only exception that Refs. [64, 71] find residual peaks in the middles of the edges of the hexagon while our calculation favors small peaks at the 120°-Néel order positions, see Fig. 7.13. As far as we can see, these small maxima decrease slightly with increasing system size and density of the frequency mesh, such that in the limit of convergence the susceptibility is possibly constant at the borders of the extended Brillouin zone. Since the response is mainly distributed along these boundary regions one may expect spin-spin correlations of very short range kind. In fact, an analysis of our real space data reveals a correlation length of $\xi \approx 0.96$ lattice spacings which agrees well with $\xi = 0.8$ found by a density matrix renormalization group (DMRG) study [64]. We may therefore conclude that the large frustration of the Kagome lattice is well captured by our FRG approach.

7.5 The Heisenberg Model on a Honeycomb Lattice

The search for quantum spin liquid phases ever since has been a complicated task. From a theorists point of view the difficulty arises from the fact that long-range correlations of any kind of operator have to be excluded. Furthermore, studies of a plethora of spin Hamiltonians on different lattices tell us that many frustrated systems tend to establish valence-bond solids which might indicate that such phases represent the generic situation in the case of magnetically disordered regimes. The remarkable numerical studies by Meng et al. [79] are a fortunate exception. In their work, they report on the first unambiguous discovery of a genuine spin-liquid phase from a generic microscopic model. They consider the Hubbard model on the honeycomb lattice by Monte Carlo

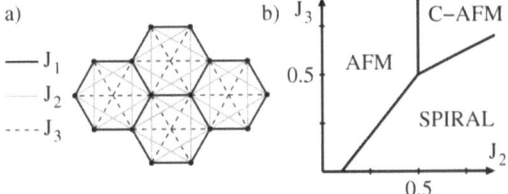

Figure 7.14: a) J_1-J_2-J_3 honeycomb model. Solid dots refer to the lattice sites. b) The classical phase diagram of the J_1-J_2-J_3 honeycomb model.

methods, which is possible for half filling where the sign problem which would generically emerge for such a system can be circumvented. At $U/t = 4.3$ they find a spin-gapped phase which shows no long-range correlations of any kind, neither charge density wave, superconductivity or even spin solid-type correlations such as that of a valence-bond solid formation. Even more importantly the study finds a clear excitation gap and no symmetry breaking of any lattice symmetry or parity P and time-reversal T, which already excludes chiral spin liquids and algebraic spin liquids.

Motivated by their work we have performed an FRG calculation for the Gutzwiller-projected version of this model, i.e., the Heisenberg model on a honeycomb lattice [96], in order to investigate which aspects of possible quantum phases may be explained through spin fluctuations only and which may necessitate the effect of charge fluctuations. The Hamiltonian is given by

$$H = J_1 \sum_{\langle i,j \rangle} \mathbf{S}_i \cdot \mathbf{S}_j + J_2 \sum_{\langle\langle i,j \rangle\rangle} \mathbf{S}_i \cdot \mathbf{S}_j + J_3 \sum_{\langle\langle\langle i,j \rangle\rangle\rangle} \mathbf{S}_i \cdot \mathbf{S}_j, \qquad (7.7)$$

where we consider interactions up to third nearest neighbors, see Fig. 7.14a, i.e., the first (second, third) sum extends over next (second, third) nearest neighbors. The phase diagram is parametrized by J_2 and J_3 given in units of J_1. The honeycomb lattice is bipartite and possesses a coordination number $z = 3$. As such, it has a lower effective dimension than the square lattice. The solution of the classical analog of the J_1-J_2-J_3 model is well known [49], see Fig. 7.14b. Being classical, it systematically orders for sufficiently small temperatures. For small J_2 the system is Néel-ordered, which is commensurate with the honeycomb lattice and preserves the sublattice 120° degrees rotational symmetry. For sufficiently low J_3 and beyond a threshold of J_2, the system minimizes the similarly important J_1 and J_2 terms and resides in a spiral phase. For high J_2 and J_3, it is energetically preferred to order in a collinear phase where spins along zigzag chains align ferromagnetically while neighboring zigzag chains exhibit antiparallel spin orientation (there are three degenerate collinear configurations).

Concerning the quantum version of the model, recent studies mainly focus on the case where $J_3 = 0$. It is well known that on this axis quantum fluctuations destroy the classical spiral order very effectively, leaving behind a (supposedly) large paramagnetic phase [38, 49, 83, 84]. However, the extent of this intermediate regime still needs to be clarified. Regarding the transition point where the Néel order melts there is remarkable disagreement: While variational Monte Carlo [38] predicts a critical point at $0.08 J_2$

7 Application to Further Models

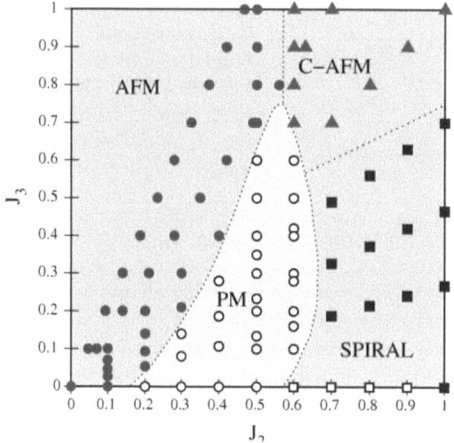

Figure 7.15: Phase diagram of the J_1-J_2-J_3 honeycomb model. The points represent parameter settings which we have computed. In the depicted J_2-J_3 range we find AFM Néel order (gray circles), collinear order (C-AFM, gray triangles), spiral order (black squares) and a paramagnetic phase (open circles). The spiral order phase partly shows incommensurability shifts (open squares) from the dominant J_2 spiral phase (see also Fig. 7.16).

(that is, much below the classical transition at $1/6J_2$), exact diagonalization [83] tends towards a value of $\sim 0.2J_2$. Moreover, the upper boundary of the paramagnetic phase where spiral order sets in, poses an even greater problem to numerics since the frustrated 120°-Néel order which exists independently on both sublattices in the limit $J_2 \to \infty$ becomes itself frustrated by a finite J_2. Up to now the literature does not contain any statement about the location of this transition. There is some body of evidence that the paramagnetic phase splits up into two regimes [38, 49, 83] where the upper one is reported to be a valence-bond solid, possibly with a staggered dimerization [83, 84]. Concerning the lower part of the intermediate phase, some works point to a spin liquid [38, 49] while others find a plaquette-valence bond solid [83]. The full J_1-J_2-J_3 model has rarely been studied and a corresponding quantum phase diagram has not yet been published. A statement about the extent of the paramagnetic phase at finite J_3 is given in Ref. [25] where the line $J_2 = J_3$ is considered. There it is found that the paramagnetic regime on this line approximately ranges from $J_2 = 0.4$ to $J_2 = 0.6$, giving way to collinear order above $J_2 = 0.6$.

For the present study our FRG algorithm internally deals with spin-spin correlations up to a length of 9 lattice constants which proved to be sufficient for a proper description of the system. Fig. 7.15 shows the quantum phase diagram of our model (7.7). In the momentum-resolved magnetic susceptibility $\chi(\mathbf{k})$, the different magnetic ordering patterns manifest as peaks in the extended Brillouin zone as depicted in Fig. 7.16. For dominant J_1 the system displays AFM order which persists longer against J_2 for finite J_3 as J_3 generally cooperates with J_1. Increasing J_2 (for not too large J_3) we clearly observe a melting of the order and the appearance of a rather large paramagnetic region. Above $J_2 \approx 0.6$ the system is characterized by presumably weak magnetic order and

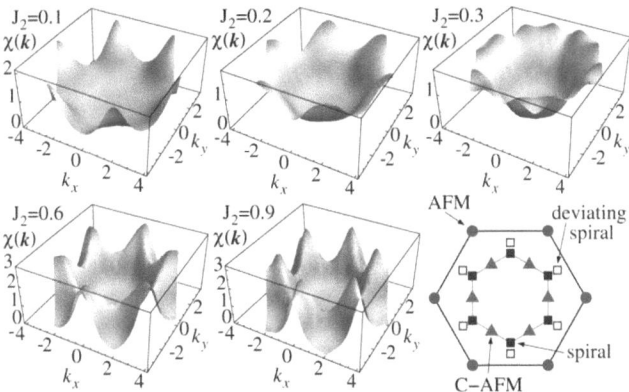

Figure 7.16: The Heisenberg-honeycomb model along the $J_3 = 0$ axis. Shown is a J_2-sweep of the static **k**-space resolved susceptibility. In magnetic phases ($J_2 = 0.1$) the susceptibility is depicted just before the instability breakdown, otherwise the physical case $\Lambda = 0$ is displayed. Susceptibilities are always given in units of $\frac{1}{J_1}$. Lower right corner: Wave-vector positions for different types of magnetic order in the extended Brillouin zone. The inner hexagon marks the first Brillouin zone.

very small ordering-instability scales which are hard to resolve numerically. However, as we enter this region by increasing J_2, at some point we observe the appearance of magnetic response peaks at discrete wave vectors **k** which we interpret as the onset of weak magnetic order. From the peak positions in **k**-space we divide this region into a collinear ordered phase for large J_3 and a spiral ordered phase for smaller J_3. The spiral phase is very close to 120°-Néel order on both sublattices of the honeycomb lattice except for a small region near the J_2-axis where the wave vector deviates from commensurability. Classically, deviations from the commensurate 120°-Néel sublattice order (full black squares in the Brillouin-zone plot in Fig. 7.16) occur throughout the spiral phase. Since quantum effects generally favor those points of high symmetry in the Brillouin zone only the classical deviations on the $J_3 = 0$ axis are strong enough to survive the effect of quantum fluctuations. We interpret this behavior as an "order by disorder" mechanism. In this respect, unlike models such as the J_1-J_2-J_3 model on a square lattice, the honeycomb model constitutes an example for a system with significant differences in the order vector between a classical and a quantum mechanical treatment. However, as also described below, the behavior on the $J_3 = 0$ axis is in agreement with the classical model and large-S calculations. Concerning the regime at large J_3, a pronounced jump of the leading susceptibility wave-vector is seen as we cross the transition between AFM and C-AFM order, pointing to a first order transition. This observation is consistent with the classical spin model except for the fact that the

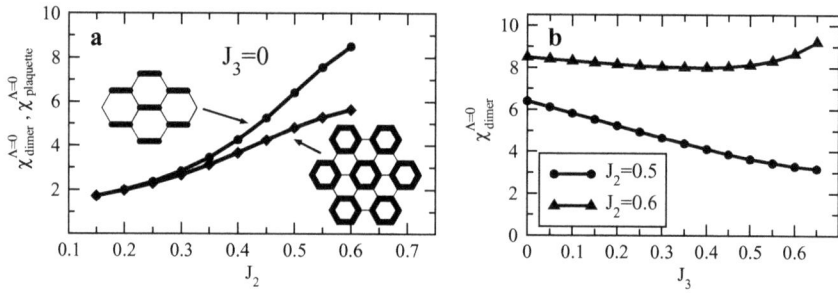

Figure 7.17: Dimer responses in the paramagnetic phase along different lines. (a) shows the staggered and plaquette responses along the $J_3 = 0$ axes and (b) the staggered response along the $J_2 = 0.5$ and $J_2 = 0.6$ lines. The dimerization patterns are depicted in (a).

transition between AFM and C-AFM order is shifted towards higher $J_{2,c} \approx 0.58$ as compared to $J_{2,c} = 0.5$ for the classical model which is due to the quantum corrections included in our calculation. On the other hand, the location of the transition line between spiral order on collinear order is not effected by quantum effects.

In order to discuss these observations in more detail, we now focus on selected lines in the quantum phase diagram. We first investigate how the fluctuation profile changes along the $J_3 = 0$ line which is depicted in Fig. 7.16. For small J_2 AFM order manifests itself in peaks at the corners of the extended Brillouin zone (see the peak positions in the schematic Brillouin-zone plot in Fig. 7.16) and the characteristic instability breakdown of the flow. As we increase J_2 the AFM peaks rapidly decrease and from the disappearance of unstable flow behavior we estimate the transition to be at $J_2 \approx 0.15$. Inside the paramagnetic phase, e.g. at $J_2 = 0.3$ no clear peak structure is visible and the susceptibility flow remains stable as we approach the physical limit $\Lambda \to 0$. Around $J_2 \approx 0.6$ spiral order peaks emerge at wave vectors slightly shifted from the commensurate positions towards larger $|\mathbf{k}|$. This feature is consistent with a large-S expansion [84] which allows to select specific wave vectors out of a classical manifold of degenerate momenta. Upon further increasing J_2, the peak positions approach commensurability, i.e., the sublattices effectively decouple and exhibit 120°-Néel order individually. During the flow, these susceptibility peaks emerge at very small Λ-scales and the instability scales are expected to be even smaller. This gives us indication of a very weak magnetization. While the transition between paramagnetism and spiral order is a bit smeared out at small J_3 the onset of response peaks occurs more abruptly at larger J_3. To resolve more information about the correlations in the disordered phase, we have computed the staggered and the plaquette dimer responses, i.e., the (dimensionless) factor of amplification of an external dimer-field perturbation exerted on system, see Eq. (6.69). With such a

7 Application to Further Models

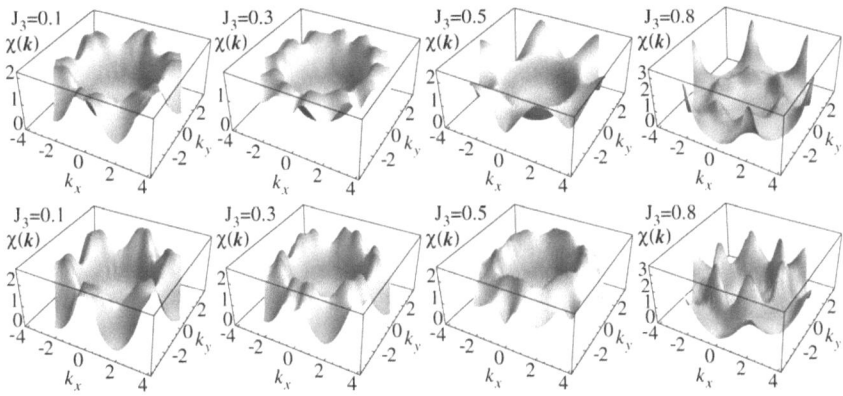

Figure 7.18: J_3-sweep of the susceptibility for $J_2 = 0.5$ (upper line) and $J_2 = 0.6$ (lower line).

perturbation-response-scheme we can distinguish between parameter regimes of different dimer fluctuation strength. For an unambiguous detection of valence-bond instabilities, however, we need to include explicit dimer susceptibilities in the form of the four-particle vertex, which is beyond the scope of our present FRG formulation. A more detailed discussion on the approximation scheme for dimer responses is contained in Section 6.6. As we sweep through the paramagnetic phase at $J_3 = 0$ we find that the staggered dimer response is dominant for higher J_2 while plaquette and staggered dimerizations compete for lower J_2, see Fig. 7.17. The absolute amplitudes are smaller for lower J_2. This is in qualitative agreement with a recent variational Monte Carlo study [38]. It supports the estimate that if at all the system will tend to form a spin liquid phase around this domain which is also the parameter regime related to the honeycomb Hubbard model from a strong coupling expansion. There, charge fluctuations which are neglected in (7.7) may be sufficient to destroy the comparably low dimer ordering tendency.

In addition, we investigate parameter lines varying J_3 for intermediate J_2 through the disordered regime. The fluctuation profiles for the paramagnetic regime at finite J_3 are shown in Fig. 7.18. There we see that at small J_3 the fluctuation profiles for $J_2 = 0.5$ and $J_2 = 0.6$ are very similar but differ more with increasing J_3 until eventually the $J_2 = 0.5$ line leads into the AFM ordered phase while collinear order emerges on the $J_2 = 0.6$ line. As we cross the transition between these two phases from $J_2 = 0.5$ to $J_2 = 0.6$, e.g. at $J_3 = 0.8$, the AFM peaks decrease while the C-AFM peaks increase such that at some point the dominant susceptibility jumps to the C-AFM wave-vector position, which characterizes the transition to be of first order. For smaller $J_3 \leq 0.6$ the paramagnetic phase shows rather complicated susceptibility profiles which have lost most signature of order fluctuations. A typical feature is the ringlike shape as seen e.g.

at $J_2 = 0.6$, $J_3 = 0.5$. An intuitive reason for a quantum disordered phase is already given from the classical limit where the point $J_2 = J_3 = 0.5$ is tricritical with three competing ordering tendencies. Concerning the staggered dimer responses along the two lines (see Fig. 7.17), with increasing J_3 there is growing indication for the formation of two regimes in the paramagnetic phase because the difference of the responses between $J_2 = 0.5$ and $J_2 = 0.6$ gets significantly larger. This leads us to the conjecture that along its full J_3-extent, the paramagnetic phase might be divided into two different phases, one being a staggered dimerized state.

In conclusion, we have studied the phase diagram of the honeycomb-Heisenberg model in the J_2-J_3 plane. We find a rather large paramagnetic phase at intermediate couplings J_2 if J_3 is not too large. Dimer responses give some indication that part of this phase (at larger J_2) might exhibit a staggered dimerization. Above $J_2 \approx 0.6$ we observe the onset of weak magnetization in the form of spiral and collinear order. A possible analog to the spin liquid found in the honeycomb-Hubbard model should emerge at smaller J_2, i.e., near the transition where the Néel order melts.

7.6 The Kitaev-Heisenberg Model

The knowledge about the ground-state properties of the spin systems considered in the last sections mainly comes from extensive numerical studies using a variety of methods. On the other hand, exact statements based on analytical calculations are rare in 2D systems. However, the exactly solvable Kitaev model [70] is a prominent exception. Its Hamiltonian consists of ferromagnetic nearest-neighbor couplings on the honeycomb lattice which are of xx, yy or zz-type, depending on the bond direction, see Fig. 7.19a. Although each of these bonds carry Ising-like interactions the ground state is non-trivial and highly frustrated. Note that the frustration does not stem from the lattice geometry or competing nearest and next-nearest neighbor interactions but rather from competing spin-anisotropy directions of the different links. In the general form of the Kitaev model the couplings corresponding to the three bond types may be different in strength. Using Majorana fermions the Hamiltonian can be reduced to a quadratic form which then allows for an exact solution [70]. In the entire parameter space spanned by the three couplings the system is in a disordered spin-liquid ground state where only nearest-neighbor spins are correlated and, moreover, for a given nearest-neighbor bond only the components of the correlator matching the bond type are non-vanishing [15]. Interestingly, these properties do not only hold for equal-time correlators but also for arbitrary time arguments. Depending on the parameter choice the system is either gapped or gapless. The Kitaev model also attracts interest because it exhibits anyonic excitations of both, Abelian and non-Abelian type. In the context of quantum computing the excitations have been proposed as robust qubits since they are protected from decoherence due to their topological nature [70].

In the following we study a certain modification of the Kitaev model. Firstly, we restrict ourselves to the case where the three couplings corresponding to the different anisotropy directions are equal. Secondly, we add further couplings to the Hamiltonian

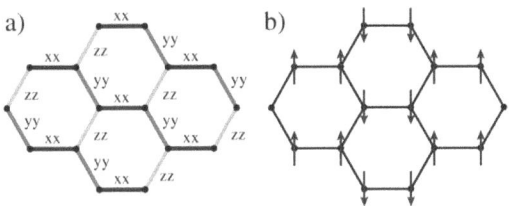

Figure 7.19: a) The Kitaev model on the honeycomb lattice. The different gray bonds correspond to the different directions of anisotropy (also denoted by xx, yy and zz). b) The stripy antiferromagnetic order.

in the form of antiferromagnetic, isotropic nearest-neighbor interactions. We call this system the Kitaev-Heisenberg model and write its Hamiltonian as

$$H = \sum_{\langle i,j \rangle} \left(J_1 \mathbf{S}_i \cdot \mathbf{S}_j - J' S_i^\gamma S_j^\gamma \right). \tag{7.8}$$

The first term represents the isotropic Heisenberg part while the second term is the Kitaev Hamiltonian with anisotropies $\gamma \equiv \gamma_{ij} = x, y, z$ depending on the bond direction. Again $\langle i, j \rangle$ denotes a sum over nearest neighbors on the honeycomb lattice. Both coupling constants are positive, $J_1, J' \geq 0$. In order to interpolate between the Néel ordered Heisenberg limit $J' = 0$ and the exactly solvable Kitaev limit $J_1 = 0$ we parametrize the exchange couplings as $J_1 = 1 - \alpha$ and $J' = 2\alpha$ and consider the whole parameter space $0 \leq \alpha \leq 1$. The phase diagram of this model is already known [34]: For small α the system is robust against anisotropic perturbations and remains in the Néel ordered state. Similarly, near the Kitaev point $\alpha = 1$ the system is stable against isotropic interactions and forms a spin-liquid phase. Remarkably a third phase, the so called stripy antiferromagnetic state, see Fig. 7.19b, emerges in the vicinity of the point $\alpha = \frac{1}{2}$. This becomes evident as one considers the system in a rotated spin basis [34]. Surprisingly, the stripy antiferromagnetic state is the exact solution at $\alpha = \frac{1}{2}$, i.e., despite of being antiferromagnetic, this point exhibits fluctuation-free classical order. Exact diagonalization [34] finds the transitions between the AF state, the stripy AF state and the spin liquid to be at $\alpha \approx 0.4$ and $\alpha \approx 0.8$, respectively. Classically, the transition between AF order and stripy AF order is located at $\alpha = \frac{1}{3}$ and is shifted upwards by quantum fluctuations [34].

Concerning experimental realizations of this model, iridium oxides such as A_2IrO_3 have been proposed as possible candidates. Ongoing experimental studies [124] indicate that Na_2IrO_3 is a Mott insulator that undergoes a transition into an AF ordered state below $T_N = 15K$. The observed antiferromagnetism already rules out that the system is near the Kitaev limit but rather in the vicinity of $\alpha = 0$. On the other hand, a large negative (antiferromagnetic) frustration parameter f, i.e., a large ratio between the Curie Weiss temperature T_{CW} and the Néel temperature T_N of $f \approx -7.7$ suggests strong frustration effects. With our FRG study presented below we intend to clarify whether these large quantum fluctuations are consistent with the Kitaev-Heisenberg model.

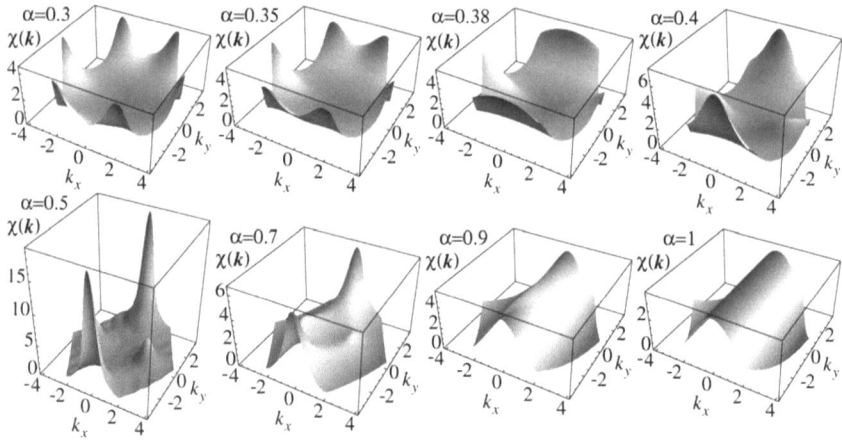

Figure 7.20: α-sweep of the static magnetic susceptibility. The response refers to a magnetic field in the anisotropy direction of a horizontal bond, i.e., according to Fig. 7.19 along the x-axis.

Aside from these experimental questions, the Kitaev-Heisenberg model is an interesting system to be studied with FRG: Firstly, the above-mentioned exact statements render it a suitable testing ground for the accuracy of our approach. Secondly, in contrast to the models considered before, it is characterized by a different frustration mechanism and (at least partially) by ferromagnetic interactions. Thirdly, it enables us to test the FRG in the case of anisotropic interactions. As a consequence of the last point, it is no longer sufficient to parametrize the two-particle vertex in the form shown in Eq. (6.36). The spin part of the two-particle vertex Γ_s splits up into components $\Gamma_{s,x}$, $\Gamma_{s,y}$ and $\Gamma_{s,z}$ such that the flow equations have to be reformulated in terms of these quantities. Apart from the fact that the resulting equations are more complex leading to longer computation times the generalization of our FRG approach to anisotropic interactions is not associated with any further difficulties.

Our FRG results for the static susceptibility are plotted in Fig. 7.20. We use the same convention for **k**-space plots as in the previous section, i.e., the depicted region is the extended (second) Brillouin zone. Due to the anisotropic interactions, the diagonal elements of the susceptibility tensor are no longer equal such that we have to specify a direction of the magnetic field for which we calculate the response (of course, we do not really apply a magnetic field to our system but calculate the susceptibility via Kubo's formula). In Fig. 7.20 we have chosen the x-direction which corresponds to the anisotropy of horizontal bonds, see Fig. 7.19. Accordingly, the susceptibility is invariant under $k_x \to -k_x$ and $k_y \to -k_y$ but has lost its 120°-rotation symmetry. Rotating the susceptibility through 120° in **k**-space is equivalent to choosing a different (cartesian) axis for the field direction. However, as seen in Fig. 7.20 even for $\alpha = 0.3$ the anisotropies have a rather small effect on the system and the instability breakdown of the flow as well as the peak structure show a clear Néel-order signature. Upon further increasing α the weight between the peaks at finite k_y rapidly rises such that at a certain value of α the wave vector in the middle between the Néel peaks becomes the leading susceptibility component. From a fine tuning of α we find this value to be $\alpha = 0.37\ldots0.38$ which we interpret as the point where the Néel order breaks down. The jump of the dominant wave vector to the new peak position occurs within a very small α-interval, indicating that in accordance with exact diagonalization [34] the transition might be of first order. Even close to criticality we find a characteristic instability breakdown that rules out the existence of an intermediate disordered phase. The peak at $\mathbf{k} = (0, \frac{2\pi}{\sqrt{3}}) \approx (0, 3.63)$ is in fact the ordering signature of the stripy AF state. Above the transition, this peak rises quickly until at $\alpha = 0.5$ (where the stripy order becomes the exact solution) the maximal height is reached. Although in the present FRG formulation we have no direct measure of the order parameter the comparably sharp peak gives some indication for strong (classical) order.

At $\alpha = 0.5$ (and partially in the plots for $\alpha = 0.4$ and $\alpha = 0.7$) another small peak is seen at wave vectors $\mathbf{k} = (\pm\frac{2\pi}{3}, 0) \approx (\pm 2.09, 0)$. In fact, it turns out that in the extended Brillouin zone the large peak alone is not sufficient to describe the stipy AF order. To motivate this property we consider the spin-spin correlation function of this state in real space: Starting from an arbitrary site, the correlation function of spins in the direction of one of the three bond orientations can be $+\frac{1}{4}$ or $-\frac{1}{4}$ (for parallel and antiparallel spin alignment) but at certain integer multiples of the nearest-neighbor lattice constant there are vacancies where the correlation can be viewed as zero. Including the vacancies the correlation function can only be fitted using two harmonics which correspond to the two different peaks. This feature is of course an artefact of our Brillouin zone convention: If we fold back the extended Brillouin zone to the first, the two peak positions coincide. In accordance with analytical considerations we obtain a ratio of the peak heights of 4.

Increasing α beyond 0.5 we observe a strong drop of the stripy AF response. Between $\alpha = 0.7$ and $\alpha = 0.9$ the peaks die out almost completely and the instability breakdown of the flow smoothly disappears, indicating the onset of paramagnetic behavior. In that regime we can only give a rough estimate of $\alpha \approx 0.7\ldots0.8$ for the transition point. Approaching the Kitaev limit $\alpha = 1$ the susceptibility becomes constant in k_y direction and has a cosine shape in k_x direction which is precisely the Fourier transform of correlated nearest-neighbor spins on x-type bonds. The exactly vanishing spin correlations beyond nearest neighbors are also seen in the real-space correlation function, demonstrating that our FRG approach is indeed able to resolve the Kitaev spin liquid. Interestingly, during the course of the Λ-flow, the value of the nearest-neighbor correlator is rather small above $\Lambda \approx 0.1$ but shows a pronounced rise below that scale which might refer to a large number of low lying states. This would be in agreement with the fact that the system is in a gapless state.

In analogy to the FRG scale Λ, also finite temperatures act as an infrared frequency cutoff. Since ordering instabilities typically occur at finite Λ-scales, it is evident already from the results using the zero-temperature FRG scheme that the Mermin-Wagner theorem cannot be fulfilled. Despite this methodological drawback, however, our results represent the generic experimental situation: Layered materials always possess small interplane couplings which generate finite (albeit small) ordering temperatures seen in experiments. On the other hand, the high-temperature physics of spin systems is governed by Curie-Weiss behavior of the uniform susceptibility,

$$\chi(\mathbf{k}=0) \propto \frac{1}{T - T_{\text{CW}}}. \tag{7.9}$$

For ferromagnetic (antiferromagnetic) materials the Curie-Weiss temperature T_{CW} is positive (negative). A comparison of the ordering temperature T_c and the Curie-Weiss temperature T_{CW} allows to determine the strength of fluctuation and frustration effects: On a classical mean-field level the frustration parameter f defined as the ratio $f = \frac{T_{\text{CW}}}{T_c}$ is one for the ferromagnet and minus one for the antiferromagnet. Switching on quantum fluctuations and/or frustrating interactions leaves the high-temperature physics such as T_{CW} unchanged but reduces the ordering temperature T_c, resulting in a frustration parameter $|f|$ larger than one. In the following we use the fact that both, the flow parameter Λ and the temperature T act as an infrared frequency cutoff. While the former is a sharp cutoff in the continuous frequency space, the latter allows a description in terms of discrete Matsubara frequencies, where the smallest mesh point sets a lower bound of the energy resolution. Even though the precise cutoff procedures associated with Λ and T are quite different, we expect that their effect is rather similar. Hence, we treat Λ like a temperature in order to be able to calculate the frustration parameter within the present FRG formulation. Indeed, it will turn out that the behavior at finite Λ and finite T is similar not only on a qualitative level but also quantitatively.

In Fig. 7.21a we present the Λ-instability scale as a function of α. A precise determination of this scale is complicated by the fact that the flow behavior near the breakdown is influenced by the discrete frequency mesh-points. The present calculation uses 46 frequencies. Those frequencies lying in the interesting Λ regime are depicted in Fig. 7.21a. It is seen that the instability scale Λ_c tends to avoid the values of the mesh. Interestingly, the transition between AF order and stripy AF order manifests itself in a pronounced kink at $\alpha = 0.37 \ldots 0.38$. Above $\alpha \approx 0.7$ the breakdown becomes smeared out which makes it difficult to observe the drop of the instability scale as we approach the spin-liquid phase.

The Curie-Weiss scale Λ_{CW} representing the physics at high energies is plotted in Fig. 7.21b. This quantity can be determined with high accuracy since over a large energy range our susceptibility data is nicely fitted by a function of the form $\sim \frac{1}{\Lambda - \Lambda_{\text{CW}}}$. The α-dependence of Λ_{CW} does not show any indication of the phase transitions but is more or less given by a linear function. At $\alpha \approx 0.7$ the Curie-Weiss scale changes sign suggesting that ferromagnetic and antiferromagnetic interactions compensate each other. Curie-Weiss behavior allows to compare the effect of Λ and T within a simple consideration: Since both, Λ_{CW} and T_{CW} describe the system at high energies, quantum fluctuations will have minor effects on these quantities. Therefore it is sufficient to compare Λ_{CW} with the classical mean-field result for T_{CW}. In fact, the curve in Fig. 7.21b is in good

7 Application to Further Models

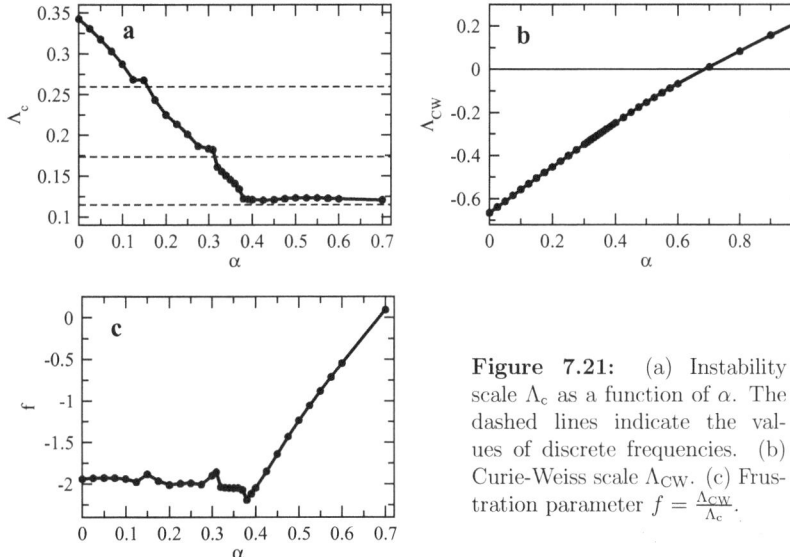

Figure 7.21: (a) Instability scale Λ_c as a function of α. The dashed lines indicate the values of discrete frequencies. (b) Curie-Weiss scale Λ_{CW}. (c) Frustration parameter $f = \frac{\Lambda_{\mathrm{CW}}}{\Lambda_c}$.

quantitative agreement with the classical result for $T_{\mathrm{CW}}(\alpha)$, which is a linear function obeying $T_{\mathrm{CW}}(0) = -0.75$ and $T_{\mathrm{CW}}(0.6) = 0$. This demonstrates that the cutoff Λ in fact behaves as a temperature and justifies our assumption that Λ can be used to obtain a finite temperature phase diagram of the Kitaev-Heisenberg model.

The frustration parameter as the ratio between both Λ-scales, $f = \frac{\Lambda_{\mathrm{CW}}}{\Lambda_c}$, is shown in Fig. 7.21c. Classically, in the AF ordered phase we would find $f = -1$ but quantum fluctuations reduce this value to $f = -2$. As soon as the stripy AF order sets in, fluctuations fall off such that at $\alpha = 0.5$ we obtain $f = -1$ which is the correct value in the case of classical order of antiferromagnetic type. The increase of f for $\alpha > 0.5$ can be assigned to the growing influence of ferromagnetic interactions. We emphasize again that a frustration parameter of $f = 0$ does not correspond to vanishing quantum fluctuations but is rather a sign of compensating ferromagnetic and antiferromagnetic interactions. Since ferromagnetic couplings dominate in the spin liquid we expect that f diverges towards plus infinity on an approach to the phase transition. As mentioned above, the critical regime is poorly resolved such that reliable data for f can only be obtained for $\alpha \lesssim 0.7$. Most importantly, however, we find that in the entire parameter regime f is by far larger than -7.7 which is the experimentally observed frustration parameter of Na_2IrO_3 [124]. Therefore, we suppose that the Kitaev-Heisenberg model is not sufficient to describe this material, possibly because interactions beyond nearest neighbors are not negligibly small or charge fluctuations play a significant role.

In summary, we find a phase diagram of the Kitaev-Heisenberg model in good agreement with Ref. [34]. Especially, the transition between AF order and stripy AF order at $\alpha = 0.37\ldots 0.38$ compares well with their result $\alpha \approx 0.4$. For the onset of the Kitaev spin liquid we can give at least a rough estimate of $\alpha = 0.7\ldots 0.8$. The classical stripy order manifests itself in a very sharp response peak and in a frustration parameter of $f = -1$. Furthermore, the Kitaev spin liquid is correctly described since spin-spin correlations vanish beyond nearest-neighbor distances. However, regarding the material Na_2IrO_3 it seems that the Kitaev-Heisenberg model cannot explain the observed strong quantum fluctuations.

8 Pseudo-Fermion FRG Including Magnetic Fields

So far we have applied our FRG scheme only in an SU(2) invariant formulation. Although we were able to detect magnetic long-range order, a controlled flow into the energy regime below the instability scale was not possible and the magnetic order parameter could not be calculated. In the present chapter we discuss if this disadvantage can be overcome within an FRG approach that explicitly includes small magnetic fields. Basically, we treat these SU(2) breaking fields as perturbations in the initial conditions and investigate their evolution as the infrared cutoff-scale Λ is lowered. In magnetically ordered phases we expect that fields matching the site dependence of the order parameter undergo a sharp increase at the critical scale and approach a finite value for $\Lambda \to 0$. The notion of symmetry breaking implies that this value does not tend to zero even if the perturbation becomes infinitesimally small. On the other hand, in disordered phases a magnetic perturbation should always result in a linear response, provided that the external fields are sufficiently small. The inclusion of magnetic fields comes along with some basic modifications of the FRG which will be presented in Section 8.1. Thereafter, in Section 8.2, as a first test of the finite-field scheme, we perform an RPA + Hartree calculation for the J_1-J_2 model where only the Hartree term of the first flow equation and the RPA term of the second flow equation is kept. Finally, Section 8.3 applies the symmetry breaking FRG scheme on the full one-loop level (i.e., including all interaction channels) and discusses the results for the J_1-J_2 model.

8.1 Modifications of the Formalism

In general, an external magnetic field modifies the parametrization of the self energy, see Eqs. (6.27) and (6.35). If we assume a field orientation in z-direction, the self energy may be written as

$$\Sigma(1,1') = \left(-i\gamma_{i_1}^\Lambda(\omega_1)\delta_{\alpha_{1'}\alpha_1} + m_{i_1}^\Lambda(\omega_1)\sigma_{\alpha_{1'}\alpha_1}^z\right)\delta_{i_1 i_{1'}}\delta(\omega_1 - \omega_{1'}), \qquad (8.1)$$

where the first term in the bracket represents the SU(2) invariant auxiliary-fermion damping while the second term includes the magnetic field m^Λ. Both, γ^Λ and m^Λ are considered as flowing quantities. The latter may have a dependence on the real space coordinate i_1 which, in principle can also generate a site dependence of the damping γ^Λ. In the most general form, m^Λ carries a frequency argument, however, the physical external field B which represents the limit $B = m^{\Lambda \to \infty}$ will be chosen as constant in

ω. Nonetheless, the frequency argument is kept since a dependence on ω is usually generated during the flow.

In the SU(2) invariant case the two-particle vertex has been parametrized in terms of Γ_s^Λ describing a spin interaction $\sim \sigma_{\alpha_{1'}\alpha_1}^\mu \sigma_{\alpha_{2'}\alpha_2}^\mu$ and Γ_d^Λ describing a density interaction $\sim \delta_{\alpha_{1'}\alpha_1}\delta_{\alpha_{2'}\alpha_2}$, see Eq. (6.36). The inclusion of a magnetic field complicates this representation considerably. In order to label the different two-particle components, we change the notation as follows: An interaction with the spin structure $\sigma_{\alpha_{1'}\alpha_1}^\mu \sigma_{\alpha_{2'}\alpha_2}^{\mu'}$ where $\mu,\mu' = 1,2,3,4$ and $\sigma^1 \equiv \sigma^x$, $\sigma^2 \equiv \sigma^y$, $\sigma^3 \equiv \sigma^z$, $\sigma^4 \equiv \delta$ is now denoted as $\Gamma_{\mu\mu'}^\Lambda$ (i.e., the relation to the old notation is given by $\Gamma_\text{s}^\Lambda \equiv \Gamma_{11}^\Lambda = \Gamma_{22}^\Lambda = \Gamma_{33}^\Lambda$ and $\Gamma_\text{d}^\Lambda \equiv \Gamma_{44}^\Lambda$). Firstly, the symmetry breaking in z-direction splits up Γ_s^Λ such that the flow of Γ_{33}^Λ is described by a separate equation, while the remaining rotation symmetry in the x-y-plane implies $\Gamma_{11}^\Lambda = \Gamma_{22}^\Lambda$. Secondly, unlike in the case of anisotropic easy-axis spin interactions where the SU(2) symmetry is broken down to Z_2 symmetry, here, we also have to deal with off-diagonal components. It turns out that Γ_{12}^Λ, Γ_{21}^Λ, Γ_{34}^Λ and Γ_{43}^Λ are finite while the other terms vanish. Using the rotation symmetry in the x-y-plane it follows that $\Gamma_{12}^\Lambda = -\Gamma_{21}^\Lambda$. We emphasize that a similar relation does not hold for Γ_{34}^Λ and Γ_{43}^Λ. In total, we have six independent flowing components, Γ_{11}^Λ, Γ_{12}^Λ, Γ_{33}^Λ, Γ_{34}^Λ, Γ_{43}^Λ and Γ_{44}^Λ which parametrize the two-particle vertex as follows,

$$\Gamma^\Lambda(1',2';1,2) = \bigg[\ \Gamma_{11\ i_1 i_2}^\Lambda(\omega_{1'},\omega_{2'};\omega_1,\omega_2)\left(\sigma_{\alpha_{1'}\alpha_1}^x \sigma_{\alpha_{2'}\alpha_2}^x + \sigma_{\alpha_{1'}\alpha_1}^y \sigma_{\alpha_{2'}\alpha_2}^y\right)$$
$$+\Gamma_{12\ i_1 i_2}^\Lambda(\omega_{1'},\omega_{2'};\omega_1,\omega_2)\left(\sigma_{\alpha_{1'}\alpha_1}^x \sigma_{\alpha_{2'}\alpha_2}^y - \sigma_{\alpha_{1'}\alpha_1}^y \sigma_{\alpha_{2'}\alpha_2}^x\right)$$
$$+\Gamma_{33\ i_1 i_2}^\Lambda(\omega_{1'},\omega_{2'};\omega_1,\omega_2)\,\sigma_{\alpha_{1'}\alpha_1}^z \sigma_{\alpha_{2'}\alpha_2}^z$$
$$+\frac{1}{i}\Gamma_{34\ i_1 i_2}^\Lambda(\omega_{1'},\omega_{2'};\omega_1,\omega_2)\,\sigma_{\alpha_{1'}\alpha_1}^z \delta_{\alpha_{2'}\alpha_2}$$
$$+\frac{1}{i}\Gamma_{43\ i_1 i_2}^\Lambda(\omega_{1'},\omega_{2'};\omega_1,\omega_2)\,\delta_{\alpha_{1'}\alpha_1}\sigma_{\alpha_{2'}\alpha_2}^z$$
$$+\Gamma_{44\ i_1 i_2}^\Lambda(\omega_{1'},\omega_{2'};\omega_1,\omega_2)\,\delta_{\alpha_{1'}\alpha_1}\delta_{\alpha_{2'}\alpha_2}\bigg]$$
$$\times \delta(\omega_{1'}+\omega_{2'}-\omega_1-\omega_2)\,\delta_{i_1 i_{1'}}\delta_{i_2 i_{2'}}$$
$$-(\omega_1 \leftrightarrow \omega_2,\,i_1 \leftrightarrow i_2,\,\alpha_1 \leftrightarrow \alpha_2)\,, \qquad (8.2)$$

Inserting Eqs. (8.1) and (8.2) into Eqs. (6.26a) and (6.26b) the flow equations can be derived straightforwardly. The Katanin truncation scheme and its extra terms are implemented in analogy to Section 6.5. Furthermore, using Eq. (6.53) the susceptibility which has longitudinal ($\chi^{\Lambda\ zz}$), transverse ($\chi^{\Lambda\ xx} = \chi^{\Lambda\ yy}$) and off-diagonal components ($\chi^{\Lambda\ xy} = -\chi^{\Lambda\ yx}$) is obtained. Compared to the SU(2) invariant case, the flow equations are much more complex and will not be presented here. The initial conditions include the external magnetic field B_i and the bare couplings $J_{i_1 i_2}$,

$$\gamma_i^{\Lambda\to\infty} = 0\,,\quad m_i^{\Lambda\to\infty} = B_i$$
$$\Gamma_{11\ i_1 i_2}^{\Lambda\to\infty} = \Gamma_{33\ i_1 i_2}^{\Lambda\to\infty} = \tfrac{1}{4}J_{i_1 i_2}\,,\quad \Gamma_{12\ i_1 i_2}^{\Lambda\to\infty} = \Gamma_{34\ i_1 i_2}^{\Lambda\to\infty} = \Gamma_{43\ i_1 i_2}^{\Lambda\to\infty} = \Gamma_{44\ i_1 i_2}^{\Lambda\to\infty} = 0\,. \qquad (8.3)$$

The central quantity of interest in this chapter, i.e., the flowing magnetization or mag-

$s \to -s,\ i_1 \leftrightarrow i_2$	$t \to -t$
$\Gamma^\Lambda_{11\,i_1 i_2}(s,t,u) = \Gamma^\Lambda_{11\,i_2 i_1}(-s,t,u)$	$\Gamma^\Lambda_{11\,i_1 i_2}(s,t,u) = \Gamma^\Lambda_{11\,i_1 i_2}(s,-t,u)$
$\Gamma^\Lambda_{12\,i_1 i_2}(s,t,u) = \Gamma^\Lambda_{12\,i_2 i_1}(-s,t,u)$	$\Gamma^\Lambda_{12\,i_1 i_2}(s,t,u) = -\Gamma^\Lambda_{12\,i_1 i_2}(s,-t,u)$
$\Gamma^\Lambda_{33\,i_1 i_2}(s,t,u) = \Gamma^\Lambda_{33\,i_2 i_1}(-s,t,u)$	$\Gamma^\Lambda_{33\,i_1 i_2}(s,t,u) = \Gamma^\Lambda_{33\,i_1 i_2}(s,-t,u)$
$\Gamma^\Lambda_{34\,i_1 i_2}(s,t,u) = -\Gamma^\Lambda_{43\,i_2 i_1}(-s,t,u)$	$\Gamma^\Lambda_{34\,i_1 i_2}(s,t,u) = \Gamma^\Lambda_{34\,i_1 i_2}(s,-t,u)$
$\Gamma^\Lambda_{44\,i_1 i_2}(s,t,u) = \Gamma^\Lambda_{44\,i_2 i_1}(-s,t,u)$	$\Gamma^\Lambda_{43\,i_1 i_2}(s,t,u) = \Gamma^\Lambda_{43\,i_1 i_2}(s,-t,u)$
	$\Gamma^\Lambda_{44\,i_1 i_2}(s,t,u) = \Gamma^\Lambda_{44\,i_1 i_2}(s,-t,u)$

$u \to -u,\ i_1 \leftrightarrow i_2$	$s \leftrightarrow u$
$\Gamma^\Lambda_{11\,i_1 i_2}(s,t,u) = \Gamma^\Lambda_{11\,i_2 i_1}(s,t,-u)$	$\Gamma^\Lambda_{11\,i_1 i_2}(s,t,u) = \Gamma^\Lambda_{11\,i_1 i_2}(u,t,s)$
$\Gamma^\Lambda_{12\,i_1 i_2}(s,t,u) = \Gamma^\Lambda_{12\,i_2 i_1}(s,t,-u)$	$\Gamma^\Lambda_{12\,i_1 i_2}(s,t,u) = \Gamma^\Lambda_{12\,i_1 i_2}(u,t,s)$
$\Gamma^\Lambda_{33\,i_1 i_2}(s,t,u) = \Gamma^\Lambda_{33\,i_2 i_1}(s,t,-u)$	$\Gamma^\Lambda_{33\,i_1 i_2}(s,t,u) = \Gamma^\Lambda_{33\,i_1 i_2}(u,t,s)$
$\Gamma^\Lambda_{34\,i_1 i_2}(s,t,u) = \Gamma^\Lambda_{43\,i_2 i_1}(s,t,-u)$	$\Gamma^\Lambda_{34\,i_1 i_2}(s,t,u) = -\Gamma^\Lambda_{34\,i_1 i_2}(u,t,s)$
$\Gamma^\Lambda_{44\,i_1 i_2}(s,t,u) = \Gamma^\Lambda_{44\,i_2 i_1}(s,t,-u)$	$\Gamma^\Lambda_{43\,i_1 i_2}(s,t,u) = \Gamma^\Lambda_{43\,i_1 i_2}(u,t,s)$
	$\Gamma^\Lambda_{44\,i_1 i_2}(s,t,u) = -\Gamma^\Lambda_{44\,i_1 i_2}(u,t,s)$

Table 8.1: Symmetries of the two-particle vertex components under frequency and site transformations.

netic order parameter M^Λ, is given by the dressed fermion bubble,

$$M_i^{\Lambda z} = \bigcirc = -\frac{1}{2}\frac{1}{2\pi}\int d\omega\, \text{Tr}[\sigma^z G_i^\Lambda(\omega)] = \frac{1}{\pi}\int_\Lambda^\infty d\omega\, \frac{m_i^\Lambda(\omega)}{(\omega + \gamma_i^\Lambda(\omega))^2 + (m_i^\Lambda(\omega))^2}. \quad (8.4)$$

Here the index i refers to the site dependence and z denotes the direction of the external field.

Before we apply the finite-field scheme to the J_1-J_2 model in the next two sections, we briefly state some properties and symmetries of the vertices which reduce the numerical effort considerably. Firstly, the quantities γ^Λ, m^Λ, Γ^Λ_{11}, Γ^Λ_{12}, Γ^Λ_{33}, Γ^Λ_{34}, Γ^Λ_{43}, Γ^Λ_{44} introduced in Eqs. (8.1) and (8.2) are defined such that they are all real. Secondly, the damping is an odd function and the magnetic field is an even function in their frequency argument, $\gamma^\Lambda(\omega) = -\gamma^\Lambda(-\omega)$ and $m^\Lambda(\omega) = m^\Lambda(-\omega)$. Interestingly, invariances of the two-particle vertex under the transformations (1) $s \to -s$, $i_1 \leftrightarrow i_2$, (2) $t \to -t$, (3) $u \to -u$, $i_1 \leftrightarrow i_2$ and (4) $s \leftrightarrow u$ may also be formulated in the SU(2) broken case. Tab. (8.1) summarizes these symmetries for each component of the two-particle vertex. There it is seen that the symmetries of Γ^Λ_s and Γ^Λ_d in the SU(2) invariant scheme remain valid. The corresponding proofs are analogous to those presented in appendix C. Thirdly, for an application to the J_1-J_2 model we are interested in field configurations with $B_i = \pm B$ where the two signs correspond to two sublattices A and B which are either of Néel or of collinear type.

8 Pseudo-Fermion FRG Including Magnetic Fields

$\Gamma^\Lambda_{11\,i_1\in A, i_2\in A} = \Gamma^\Lambda_{11\,i_{1'}\in B, i_{2'}\in B}$	$\Gamma^\Lambda_{11\,i_1\in A, i_2\in B} = \Gamma^\Lambda_{11\,i_{1'}\in B, i_{2'}\in A}$
$\Gamma^\Lambda_{12\,i_1\in A, i_2\in A} = -\Gamma^\Lambda_{12\,i_{1'}\in B, i_{2'}\in B}$	$\Gamma^\Lambda_{12\,i_1\in A, i_2\in B} = -\Gamma^\Lambda_{12\,i_{1'}\in B, i_{2'}\in A}$
$\Gamma^\Lambda_{33\,i_1\in A, i_2\in A} = \Gamma^\Lambda_{33\,i_{1'}\in B, i_{2'}\in B}$	$\Gamma^\Lambda_{33\,i_1\in A, i_2\in B} = \Gamma^\Lambda_{33\,i_{1'}\in B, i_{2'}\in A}$
$\Gamma^\Lambda_{34\,i_1\in A, i_2\in A} = -\Gamma^\Lambda_{34\,i_{1'}\in B, i_{2'}\in B}$	$\Gamma^\Lambda_{34\,i_1\in A, i_2\in B} = -\Gamma^\Lambda_{34\,i_{1'}\in B, i_{2'}\in A}$
$\Gamma^\Lambda_{43\,i_1\in A, i_2\in A} = -\Gamma^\Lambda_{43\,i_{1'}\in B, i_{2'}\in B}$	$\Gamma^\Lambda_{43\,i_1\in A, i_2\in B} = -\Gamma^\Lambda_{43\,i_{1'}\in B, i_{2'}\in A}$
$\Gamma^\Lambda_{44\,i_1\in A, i_2\in A} = \Gamma^\Lambda_{44\,i_{1'}\in B, i_{2'}\in B}$	$\Gamma^\Lambda_{44\,i_1\in A, i_2\in B} = \Gamma^\Lambda_{44\,i_{1'}\in B, i_{2'}\in A}$

Table 8.2: Symmetries of the two-particle vertex components under lattice translations. A and B denote two equivalent sublattices. The relations are valid if $\mathbf{R}_{i_1} - \mathbf{R}_{i_2} = \mathbf{R}_{i_{1'}} - \mathbf{R}_{i_{2'}}$. All frequency arguments are omitted.

Due to the equivalence of the sublattices the relations $\gamma^\Lambda_{i\in A} = \gamma^\Lambda_{i\in B}$ and $m^\Lambda_{i\in A} = -m^\Lambda_{i\in B}$ hold. Similar identities are found for the two-particle vertex, see Tab. (8.2). After these methodological considerations we now present applications to the J_1-J_2 model.

8.2 Hartree- and Random Phase Approximation

Before we turn to the full one-loop magnetic FRG scheme in the next section, we briefly discuss the special case where only the Hartree (Fig. 6.3a) and RPA (Fig. 6.2b) terms are kept in the first and second flow equation, respectively. As demonstrated in Section 6.5, in conjunction with the Katanin truncation such a scheme is equivalent to the conserving RPA + Hartree approximation. This equivalence has already been shown in the paramagnetic phase of the J_1-J_2 model, see Fig. 6.6. However, without SU(2) symmetry breaking fields the magnetic phases have not been accessible. Using the formalism sketched in the last section this can now be accomplished. In this respect, our approach resembles the studies in Refs. [52, 107] where symmetry broken FRG flows into charge-density wave phases and superconducting phases are considered. In order to model the effect of quantum fluctuations and to make the solution less trivial we again include a phenomenological pseudo-fermion damping γ as defined in Eqs. (5.5) and (5.6) which is kept constant during the flow.

The restriction to the RPA and Hartree channels is accompanied with enormous simplifications of the flow equations. Most importantly, the FRG flow does not generate dependences on the transfer frequencies s and u. Similarly, as long as the external magnetic field $B = m^{\Lambda\to\infty}$ is constant in ω, it remains constant during the flow. On the other hand, the two-particle vertex carries a dependence on t, however, flow equations corresponding to different t-arguments are decoupled such that the static component may be considered individually. Moreover, the spin components of the two-particle vertex are independent of each other and it turns out that only Γ^Λ_{11} and Γ^Λ_{33} are non-zero. Here, we concentrate on the flow of Γ^Λ_{33}, from which the longitudinal susceptibility $\chi^{\Lambda\,zz}$ is obtained. Taking into account all these simplifications one ends up with the following

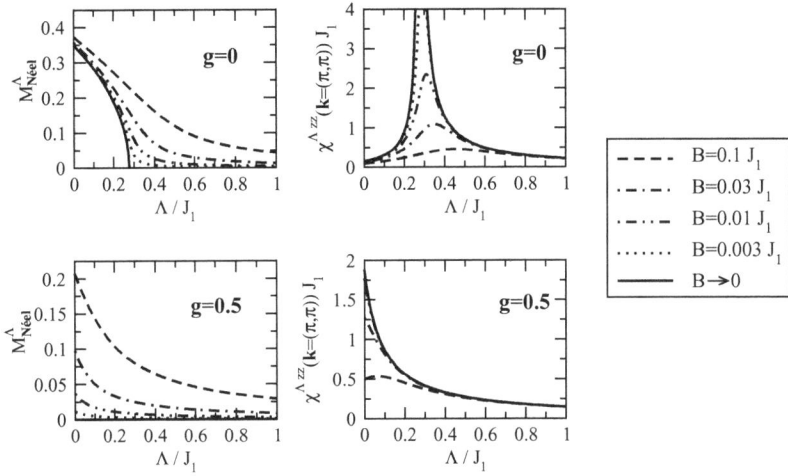

Figure 8.1: RPA + Hartree approximation for the J_1-J_2 model within FRG: Néel magnetization $M^\Lambda = |M_i^\Lambda|$ and longitudinal Néel susceptibility $\chi^{\Lambda\,zz}$ as a function of Λ for various values of the external magnetic field $B = |B_i|$. The damping is chosen according to Eq. (5.6) with $\tilde{\gamma} = 0.36$. The upper row shows the flow into the Néel ordered phase at $g = 0$ while the lower row represents the paramagnetic phase at $g = 0.55$.

equations for the flowing magnetic field m_i^Λ, the longitudinal two-particle vertex $\Gamma_{33\,i_1 i_2}^\Lambda$, the local magnetic order parameter M_i^Λ and the longitudinal spin-spin correlator $\chi_{i_1 i_2}^{\Lambda\,zz}$,

$$\frac{d}{d\Lambda} m_{i_1}^\Lambda = \frac{2}{\pi} \sum_{i_2} \Gamma_{33\,i_1 i_2}^\Lambda \frac{m_{i_2}^\Lambda}{(\Lambda+\gamma)^2 + (m_{i_2}^\Lambda)^2}\,, \tag{8.5}$$

$$\frac{d}{d\Lambda} \Gamma_{33\,i_1 i_2}^\Lambda = \frac{2}{\pi} \sum_j \Gamma_{33\,i_1 j}^\Lambda \Gamma_{33\,j i_2}^\Lambda \frac{(\Lambda+\gamma)^2 - (m_j^\Lambda)^2 + 2(\Lambda+\gamma) m_j^\Lambda \frac{d}{d\Lambda} m_j^\Lambda}{((\Lambda+\gamma)^2 + (m_j^\Lambda)^2)^2}\,, \tag{8.6}$$

$$M_i^\Lambda = \frac{1}{\pi} \arctan\left(\frac{m_i^\Lambda}{\Lambda+\gamma}\right)\,, \tag{8.7}$$

$$\chi_{i_1 i_2}^{\Lambda\,zz} = \frac{1}{2\pi} \frac{\Lambda+\gamma}{(\Lambda+\gamma)^2 + (m_{i_1}^\Lambda)^2} \delta_{i_1 i_2} - \frac{1}{\pi^2} \Gamma_{33\,i_1 i_2}^\Lambda \frac{(\Lambda+\gamma)^2}{((\Lambda+\gamma)^2 + (m_{i_1}^\Lambda)^2)((\Lambda+\gamma)^2 + (m_{i_2}^\Lambda)^2)}\,. \tag{8.8}$$

Note that all quantities represent static components. The initial conditions for m_i^Λ and $\Gamma_{33\,i_1 i_2}^\Lambda$ have already been stated in Eq. (8.3). These equations are most effectively solved in **k**-space where the flow of different wave-vector components decouples. We assume a Néel-like modulation of the magnetic field B_i and calculate the response in the form of the Néel magnetization $M^\Lambda = |M_i^\Lambda| \in [0, 0.5]$ and the staggered longitudinal susceptibility $\chi^{\Lambda\,zz}(\mathbf{k} = (\pi, \pi))$. The damping parameter $\tilde{\gamma} = 0.36$ (see Eq. (5.6)) is chosen such that the paramagnetic phase has the correct extent.

Fig. 8.1 shows the results of the calculation. It is seen that in the Néel phase at $g = 0$ the magnetic moment M behaves as if Λ was a temperature: With decreasing magnetic field the rise at $\Lambda \approx 0.28$, signalling the symmetry breaking, gets sharper and the magnetization at zero Λ-scale converges towards the mean-field order parameter shown in Fig. 5.1. Regarding the longitudinal susceptibility, the divergence occurring at zero field is regularized as the magnetic field is switched on. The peak height is proportional to the reciprocal of B and the peak position is shifted towards larger Λ with increasing field. A finite coupling J_2 lowers the instability scale and weakens the magnetization. Finally, when the paramagnetic regime is reached (see the plots for $g = 0.5$) the B-field perturbation generates a linear response, i.e., the magnetization vanishes linearly with the magnetic field and the susceptibility rapidly converges towards the zero-field limit.

8.3 Full One-Loop FRG Scheme

In this section we tackle the full magnetic FRG scheme considering all one-loop interaction channels as well as the Katanin scheme. Due to the additional non-vanishing spin components of the two-particle vertex and the more complicated structure of the flow equations, the solution requires a lot of computational effort. In the case that a Néel-like external magnetic field is imposed on the J_1-J_2 Heisenberg system the computing times are approximately 18 times longer compared to the SU(2) invariant scheme. This becomes even worse if we apply a magnetic field with a collinear ordering pattern: The broken 90°-rotation symmetry results in an additional factor of two in the computation times. Hence, we will not consider collinear order in the following but restrict ourselves to Néel-like perturbations. In our calculations we use 46 frequency mesh-points and include correlations up to a length of 5 lattice spacings.

The results of such calculations are shown in Fig. 8.2. For $g = 0$ we obtain a similar behavior as compared to the RPA + Hartree scheme of the last section: The rise of the magnetization becomes more and more pronounced as the magnetic field is lowered. Generally, we are not able to apply arbitrarily small external fields. For $B \lesssim 0.01 J_1$ we could not reach the intersting Λ-regime below $\Lambda \approx 0.5$ due to a sudden breakdown of the flow. This may be explained by the fact that the energy scale given by the field falls below the lowest discrete frequency mesh-point. For the smallest accessible fields we obtain a magnetization of $M^{\Lambda \to \infty} \approx 0.42$, i.e., clearly larger than $M = 0.3$ found by Monte Carlo [94]. Regarding the longitudinal susceptibility the instability breakdown that would occur at $\Lambda \approx 0.45 J_1$ is regularized by a finite field such that we do not observe unstable oscillatory flow behavior below that scale. Altogether, our results at finite fields clearly support the notion of Néel order in the nearest-neighbor Heisenberg model.

However, the flow behavior becomes ambiguous if we consider the paramagnetic regime

8 Pseudo-Fermion FRG Including Magnetic Fields

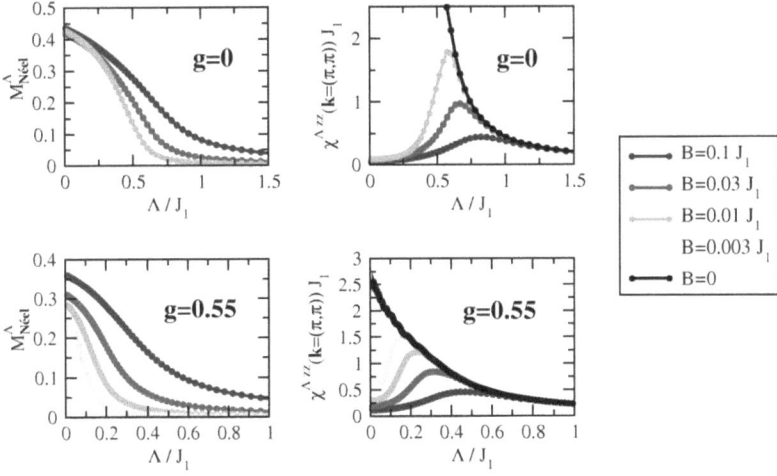

Figure 8.2: Magnetic FRG scheme including all interaction channels: Néel magnetization $M^\Lambda = |M_i^\Lambda|$ and longitudinal Néel susceptibility $\chi^{\Lambda\,zz}$ as a function of Λ for various values of the external magnetic field $B = |B_i|$. The upper row shows the flow into the Néel ordered phase at $g = 0$ while the lower row represents the paramagnetic phase at $g = 0.55$.

at $g = 0.55$. Even at rather small fields we obtain large magnetizations that would point to the existence of Néel order. (For the smallest field $B = 0.003 J_1$ a stable flow could only be obtained above $\Lambda \approx 0.05 J_1$.) By contrast, the susceptibility data at $g = 0.55$ implies that at small enough magnetic fields the curves converge towards the SU(2) invariant solution, which would correspond to a linear response. Unfortunately, the numerically accessible settings for the frequency discretization and the external field do not allow to clarify this issue.

We emphasize that except for the case of a conserving approximation (which fulfills Ward identities exactly) the phase boundaries detected by the susceptibility at zero field are not necessarily identical to those found by the onset of non-linear response to a small perturbation. Within a conserving scheme the susceptibility diagrams are generated by the derivative of magnetization diagrams (bubbles of a single fermion propagator) with respect to the magnetic field. This assures that the two methods of detecting phase transitions yield the same results. However, if one spoils Ward identities as we do, susceptibility diagrams obtained by the derivative of the magnetization are not identical to those generated during the RG flow, leading to ambiguities in the phase diagram when a magnetic field is applied. We suppose that this is the origin of the unclear flow behavior in the paramagnetic phase.

In conclusion, an FRG scheme reproducing the results of an RPA + Hartree approximation demonstrates that in view of the flow behavior of the magnetization and the susceptibility, the cutoff Λ may be regarded as a temperature. Our findings within the RPA + Hartree scheme agree qualitatively well as compared to the full one-loop calculation, at least in ordered phases. However, the intermediate regime is sensitive to external Néel perturbations such that the magnetic properties at $g = 0.55$ under the influence of small fields remain ambiguous. Therefore, we conclude that within an FRG scheme that does not fulfill Ward identities exactly the response to small magnetic fields is not a good criterion to identify different phases of the J_1-J_2 model.

9 FRG at Finite Temperatures

As the FRG is typically formulated in the framework of the Matsubara technique, it is of course not restricted to zero temperature. In this chapter, we first present the modifications of our FRG approach which are necessary to allow for finite temperatures. Secondly, we apply this scheme to the J_1-J_2 model and discuss the results. In particular, since the chemical potential used within the Popov scheme $\mu^{\text{ppv}} = -\frac{i\pi T}{2}$ becomes finite, it can be explicitly included in our numerics which enables us to test the fulfillment of the pseudo-particle constraint.

9.1 Modifications of the Formalism

The most obvious consequence of finite temperatures is that the Matsubara frequencies are discrete, i.e., $\omega_n = (2n+1)\pi T$ for fermionic and $\nu_n = 2n\pi T$ for bosonic frequencies ($n \in \mathbb{Z}$). This means that the internal integrations become sums. Since only a finite number of these frequencies may be treated numerically, we have to choose a set of indices n_i. Most naturally, below some index n_0 all frequencies are included while with increasing $|n| > n_0$ more and more integers are omitted. In order to generate such a set we use a function of the form

$$n_i = \text{round}\left(z \sinh\left(\frac{i}{z}\right)\right) \quad \text{with} \quad i = -i_0, -i_0+1, \ldots, i_0, \tag{9.1}$$

where the parameter z determines the value of n_0 (for a symmetric set of fermionic Matsubara frequencies we have to use a slightly different formula). For the evaluation of vertex functions at frequency arguments that do not correspond to integers included in the set, we perform a linear interpolation between adjacent mesh points as described in Section 6.2.

The legs of the m-particle vertex functions all carry fermionic frequencies, implying that the transfer frequencies of the two-particle vertex s, t, u are bosonic. If we write the frequencies on the legs of the two-particle vertex as $\omega_x = (2n_{\omega_x}+1)\pi T$ with $x = 1, 1', 2, 2'$ and the transfer frequencies as $y = 2n_y\pi T$ with $y = s, t, u$ then the relations in Eq. (6.39) transform into the following rules

$$n_s = n_{\omega_{1'}} + n_{\omega_{2'}} + 1, \quad n_t = n_{\omega_{1'}} - n_{\omega_1}, \quad n_u = n_{\omega_{1'}} - n_{\omega_2}. \tag{9.2}$$

These equations have an interesting implication: While n_{ω_1}, $n_{\omega_{1'}}$, n_{ω_2} take all integer values independently ($n_{\omega_{2'}}$ is fixed by energy conservation), the numbers n_s, n_t, n_u are dependent of each other. It turns out that either one or three of them are odd while the others are even. In other words, the possible values for n_s, n_t, n_u form a face-centered cubic lattice. If we now thin out this lattice at higher frequencies in order to obtain a numerically treatable mesh, we are faced with the non-trivial problem of finding a scheme that interpolates between the remaining points. However, we avoid this difficulty by parametrizing the two-particle vertex by n_{ω_1}, $n_{\omega_{1'}}$, n_{ω_2} instead of n_s, n_t, n_u. Since n_{ω_1}, $n_{\omega_{1'}}$, n_{ω_2} form a simple cubic lattice, the linear interpolation is straightforwardly performed using Eq. (6.49).

At least within the average projection scheme where $\mu = 0$, the flow equations at finite temperatures are again invariant under the transformations $n_s \to -n_s$, $n_t \to -n_t$, $n_u \to -n_u$ and $n_s \leftrightarrow n_u$. Despite our new parametrization of the two-particle vertex we employ these symmetries and restrict the flow equations to frequency indices n_{ω_1}, $n_{\omega_{1'}}$, n_{ω_2} that correspond to positive n_s, n_t, n_u. (In the case that the right side of a flow equation exhibits a vertex at frequencies belonging to negative n_s, n_t, n_u we identify it with the vertex carrying the transfer frequencies $|n_s|$, $|n_t|$, $|n_u|$.) Note that the Popov scheme effectively shifts all Matsubara frequencies by one quarter of their distance such that they are no longer symmetric around zero. Hence, the above-mentioned symmetries are all violated and we have to deal with the full (depleted) mesh spanned by n_{ω_1}, $n_{\omega_{1'}}$, n_{ω_2}, resulting in an additional factor of 16 in the computation times.

A further modification concerns the infrared frequency cutoff: The sharp function $\Theta(|\omega_n| - \Lambda)$ is not compatible with discrete Matsubara sums $\sum_{\omega_n}$ since the single-scale propagator $S^\Lambda \propto \delta(|\omega_n| - \Lambda)$ gives an (infinite) contribution only if the scale Λ matches a Matsubara frequency. Hence, a sharp cutoff does not generate continuous RG flows. In the following we use a cutoff function of the form

$$\chi^\Lambda(\omega_n) = \frac{\omega_n^2}{\omega_n^2 + \Lambda^2}, \qquad (9.3)$$

see e.g. Ref. [59], with a width of the "step" at $\omega_n \approx \pm\Lambda$ that increases with Λ. We emphasize that there is no unique choice for this function. In fact, many other possibilities are found in the literature [58, 66]. Using $\chi^\Lambda(\omega_n)$ as defined in Eq. (9.3) the Green's function and the single-scale propagator are given by

$$G^\Lambda(\omega_n) = \frac{1}{i} \frac{\omega_n \eta}{\omega_n^2 + \omega_n \gamma^\Lambda(\omega_n)\eta + \Lambda^2}, \qquad (9.4)$$

$$S^\Lambda(\omega_n) = \frac{1}{i} \frac{1}{(\omega_n^2 + \omega_n \gamma^\Lambda(\omega_n)\eta + \Lambda^2)^2} \left(2\Lambda \omega_n \eta + \omega_n^2 \eta^2 \frac{d}{d\Lambda} \gamma^\Lambda(\omega_n) \right), \qquad (9.5)$$

where η is defined as $\eta = 1$ for the average projection scheme and $\eta = 1/(1 - \frac{\pi T}{2\omega_n})$ for the Popov scheme. The term with the derivative of γ^Λ is the Katanin part of the single-scale propagator. Due to the broadened cutoff, both terms in the single-scale propagator require a frequency summation. Taking into account all these modifications, our computer code has to be changed fundamentally as compared to the zero-temperature scheme.

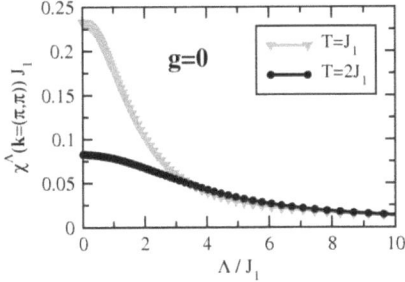

Figure 9.1: Néel-susceptibility flow within the finite temperature FRG scheme at the nearest-neighbor Heisenberg point $g = 0$ using the average projection with $\mu = 0$.

9.2 Results for the J_1-J_2 model

To begin with, we discuss the typical susceptibility flow-behavior at finite temperatures. As an example Fig. 9.1 depicts the Néel susceptibility at $g = 0$ within the average projection scheme for $T = J_1$ and $T = 2J_1$. The qualitative behavior is not changed by the Popov scheme or different choices of the frustration g. The data show an inflection point at a Λ-scale proportional to the temperature. At this scale the temperature becomes the relevant energy cutoff in the system such that with decreasing Λ the gradient falls off. Furthermore, it is seen that at these (rather large) temperatures the Néel instability is entirely regularized, leaving no sign of any oscillatory flow behavior down to the smallest scales.

In order to discuss the temperature dependence of the susceptibility, Fig. 9.2 shows the results in the physical limit $\Lambda \to 0$ and compares the two projection schemes. In fact, we cannot reach arbitrarily small temperatures for the following reason: As Λ is lowered, the cutoff gets sharper such that the internal Matsubara sum runs over a single-scale propagator with a small peak width. For a proper evaluation of this sum our mesh needs to include all Matsubara frequencies at that scale. Therefore, at small temperatures, well converged results require many frequency mesh-points (it is tempting to avoid this problem by choosing a cutoff function with a constant width of the step but this leads to an insufficient regularization of the propagators in the infrared limit).

For $g = 0$ we obtain converged results down to temperatures $T \approx 0.25 J_1$ while below the effects of the finite frequency grid and especially of the finite system size are too strong. However, at the smallest accessible temperatures a steep increase of the susceptibility is seen, which is in agreement with the existence of Néel order in that parameter regime. By contrast, at $g = 0.55$, smaller finite-size effects enable us to reach temperatures of the order of $T \approx 0.05 J_1$, at least within the average projection (calculations within the Popov scheme at such small temperatures require too much numerical resources). We stress that in general, different cutoff functions do not necessarily yield the same solution in the limit $\Lambda \to 0$ which is why we cannot expect to reproduce our zero-temperature results within the present scheme. Nevertheless, on the basis of our susceptibility data, it seems that with decreasing temperature our ground-state result, marked with a black diamond, is reached.

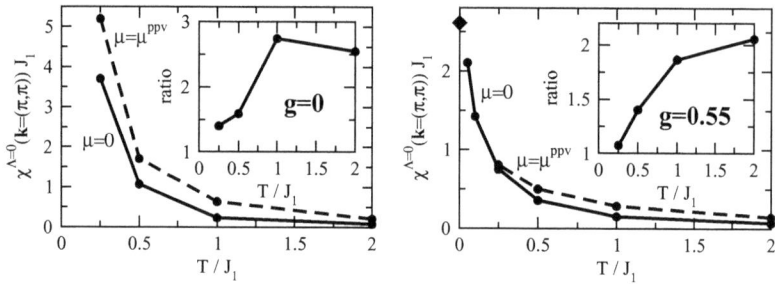

Figure 9.2: Néel susceptibility in the physical limit $\Lambda \to 0$ as a function of the temperature using the average ($\mu = 0$, full line) and the exact projection scheme ($\mu = \mu^{\mathrm{ppv}} = -\frac{i\pi T}{2}$, dashed line). The insets show the ratios of the susceptibilities for $\mu = \mu^{\mathrm{ppv}}$ and $\mu = 0$. The left plot represents the Néel phase at $g = 0$ and the right plot the paramagnetic phase at $g = 0.55$. The black diamond on the $T = 0$-axis of the right plot marks the solution of the zero temperature FRG scheme.

As a generic feature we obtain smaller susceptibilities within the average projection as compared to the Popov scheme. This can be explained by the fact that at non-zero temperatures the average scheme generally allows for sites with unphysical auxiliary-particle occupations. Since these sites do not carry spins they reduce the magnetic susceptibility. The insets show the ratio of the susceptibilities obtained within both projection schemes. It is seen that with decreasing temperature this ratio approaches one, which supports our argument that in the zero-temperature limit the average projection is sufficient to fulfill the constraint exactly.

In order to detect Curie-Weiss behavior of the homogeneous susceptibility, more data points are needed. As shown in Section 7.6, the physics at high energies such as Curie-Weiss behavior is most effectively studied on the basis of the Λ-flow within the zero-temperature scheme, because only one calculation is necessary to scan all energy scales. In view of our finite-temperature data we can at least say that with increasing T the fermion-bubble contribution to the susceptibility, i.e., the first term on the right side of Eq. (6.52), becomes dominant as compared to the remaining two-particle vertex term. This means that the system behaves more and more like a non-interacting system. In turn, with increasing T the influence of the self energy on the fermion bubble becomes irrelevant such that in the limit $T \to 0$ the susceptibility effectively reduces to the bare (unrenormalized) bubble Π_0, which can be easily calculated,

$$\Pi_0(\omega = 0) = \begin{cases} \frac{1}{8T} & \text{for } \mu = 0 \\ \frac{1}{4T} & \text{for } \mu = \mu^{\mathrm{ppv}} \end{cases}. \tag{9.6}$$

This is obviously the contribution that generates the Curie-$\frac{1}{T}$-behavior. From this equa-

9 FRG at Finite Temperatures

tion it is also seen that the susceptibility ratio of the two projection schemes converges towards two in the limit $T \to \infty$.

In conclusion, the finite temperature FRG scheme provides further indication for the existence of Néel order at $g = 0$ and magnetic disorder at $g = 0.55$. However, since small temperatures are out of the reach of the present formulation, for an unambiguous detection of different phases the ground-state FRG scheme is more suitable. Concerning the auxiliary-particle constraint our results clearly support the argument that the average projection is sufficient for an exact fulfillment at zero temperature.

10 Limitations of the FRG: Lower Dimensions

While our results for the two-dimensional spin models presented in Chapters 6 and 7 are in good agreement with other studies, real numerical tests of our approach aside from the determination of critical couplings could not be performed because almost no exact results are available in two dimensions. By contrast, a lot of analytical statements exist for one- (or even zero-) dimensional models and, moreover, these systems can be studied within shorter computation times. However, it seems to be a generic property of the spin FRG that magnetic ordering tendencies are overestimated in lower dimensions. This will be exemplified in the following.

A one-dimensional spin system containing interesting physics is the J_1-J_2 Heisenberg model on a chain with nearest-neighbor interactions J_1 and next-nearest neighbor interactions J_2. Below a critical coupling $J_{2,c} = 0.2411 J_1$ the system features a spin liquid with algebraically decaying correlations, while above the system is dimerized, gapped and short-range correlated [88]. The point $\frac{J_2}{J_1} = 0.5$, also known as the Majumdar-Ghosh point, is special because its ground state has a simple form: It is the product of spin singlets between nearest-neighbor sites, i.e., the fully dimerized state is the exact solution. Within our FRG approach we obtain the susceptibilities shown in Fig. 10.1. It is seen that in the nearest-neighbor limit $J_2 = 0$ the Néel susceptibility features an unstable, oscillatory behavior below $\Lambda \approx 0.3 J_1$, which would imply magnetic long-range order. Indeed, the breakdown is not as pronounced as in the two-dimensional case, see Fig. 6.12 but would still be consistent with a small magnetization. As J_2 is increased towards the Majumdar-Ghosh point, the susceptibility does not undergo any significant change (except for the fact that the maximum in k-space moves away from $k = \pi$). At $\frac{J_2}{J_1} = 0.5$ the wiggly behavior is smoother but still visible. Especially, regarding the spin correlations, which are exponentially decaying throughout the parameter space, we cannot resolve the transition at $\frac{J_2}{J_1} = 0.2411$. Altogether, the changes in the susceptibility upon variation of the couplings are not as pronounced as in two dimensions.

Next, we go even further and consider the two-site Heisenberg molecule, $H = J\mathbf{S}_1 \mathbf{S}_2$, as an example for a system with a finite Hilbert space. Diagonalizing the Hamiltonian, it is easily found that for $J > 0$ the ground state is a spin singlet with a static spin-spin correlator $\chi_{ij} = \langle\langle S_i^z S_j^z \rangle\rangle$ of $\chi_{11} = \chi_{22} = \frac{1}{2J}$ and $\chi_{12} = \chi_{21} = -\frac{1}{2J}$. Regarding these values the FRG provides reasonable results, i.e., $\chi_{11}^{\text{FRG}} \approx 0.61\frac{1}{J}$, $\chi_{12}^{\text{FRG}} \approx -0.36\frac{1}{J}$ and, in particular, a smooth flow behavior down to $\Lambda \to 0$ is obtained. However, the overestimation of magnetic order is again seen: Within FRG the auxiliary-particle damping $\gamma^{\Lambda \to 0}(\omega)$ comes out much smaller than the exact solution $\gamma(\omega) = \frac{9J^2}{16\omega}$, especially at small frequencies where the FRG result approaches zero while the exact solution

10 Limitations of the FRG: Lower Dimensions

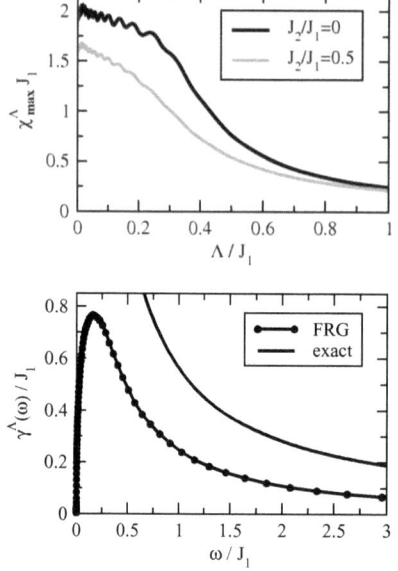

Figure 10.1: Flowing susceptibility for the one-dimensional J_1-J_2 Heisenberg model at $J_2/J_1 = 0$ and $J_2/J_1 = 0.5$. Shown is the maximum of the susceptibility in k-space, i.e., the corresponding momenta are $k = \pi$ for $J_2/J_1 = 0$ and $k \approx 2.3$ for $J_2/J_1 = 0.5$.

Figure 10.2: Auxiliary particle damping $\gamma^\Lambda(\omega)$ in the limit $\Lambda \to 0$ for the two-site Heisenberg molecule within FRG in comparison with the exact result $\gamma(\omega) = \frac{9J^2}{16\omega}$.

diverges, see Fig. 10.2. As demonstrated in Chapter 5 the damping mimics quantum fluctuations such that the system is biased towards magnetic order if γ is too small.

Although the lack of quantum fluctuations is not sufficient to generate an instability breakdown during the flow of the two-site molecule, its effect can be explicitly seen in related spin systems. We studied the bilayer-Heisenberg model that possesses nearest-neighbor couplings J_1 within the square-lattice planes and interlayer couplings $J_\perp$ that connect neighboring sites between the layers. Starting from the isolated plane limit $J_1 = 0$, with increasing $J_\perp$, the formation of interlayer spin singlets becomes more and more favored until at a certain point the Néel order breaks down. Numerical studies, in particular quantum Monte Carlo, which is not affected by the sign problem, determine the critical point with high accuracy. They find $J_{\perp,c} \approx 2.5 J_1$ [137, 139]. By contrast, the FRG still shows a clear Néel instability at $\frac{J_\perp}{J_1} = 2.5$ because the pseudo-fermion damping generated by the bond formation is too weak to destroy the order. Instead, FRG predicts the transition to be at $\frac{J_\perp}{J_1} \approx 4.5\ldots 5$.

The systematic overestimation of magnetic order in dimensions lower than two can probably be traced back to the graphs included in the FRG. The only diagrammatic contribution that is able to describe correlations of arbitrary distance is the RPA. However, this term alone corresponds to the classical mean-field theory which is known to give reasonable results at high enough dimensions but overestimates magnetic order in lower dimensions. This means that in 1D and 0D the balance between diagrams favoring order and those favoring disorder is no longer guaranteed. Therefore, we suppose that in order to describe the physics of spin chains or point-like objects correctly, further "long-range diagrams" of two-loop type have to be included. Of course, this poses a severe problem to numerics and will not be considered here.

To summarize, in this chapter we illustrated that due to the restriction to one-loop diagrams (with the exception of the Katanin terms), the spin FRG tends to overestimate magnetic order in dimensions lower than two. This problem also occurs when objects of small dimensionality are imbedded in a system of higher dimension, as in the case of the bilayer-Heisenberg model. Since, on the other hand, the investigation of 3D systems requires a lot of computational resources, the real power of our approach lies in two dimensions.

11 Conclusion and Outlook

The aim of this work is the development of new methods to calculate magnetic properties of frustrated two-dimensional quantum spin models. The starting point of our approach is perturbation theory in the exchange couplings, summed to infinite order. In order to be able to use standard many-body techniques and to perform diagrammatic expansions, we have applied the auxiliary-fermion representation of spin operators. To enforce the auxiliary-particle constraint, two different projection schemes have been employed: (i) enforcement of the constraint on the average (which, however, becomes exact at zero temperature), and (ii) exact projection using an imaginary chemical potential [91].

In a first exploratory study we use the bare RPA + Hartree approximation to access the ground-state properties. We find that on a simple mean-field level the suppression of magnetic order as an effect of frustration cannot be described. Therefore, we introduce a phenomenological damping term in the bare pseudo-fermion Green's function to account for scattering processes of the fermions which leads to a finite lifetime and a spectral broadening. We show that this damping can be tuned such that the magnetization, the susceptibility and hence the overall phase diagram of a frustrated quantum spin model come out as expected, which is demonstrated in Chapter 5 using the J_1-J_2 model. Hence, as an important result for the subsequent chapter we find that the pseudo-fermion damping is the central quantity to control the competition between order and disorder fluctuations. However, it is demonstrated that the microscopic derivation of the pseudo-fermion damping is beyond the reach of simple diagrammatic resummations.

A more systematic approach is considered in the main part of the thesis: the functional renormalization group (FRG) method. In Chapter 6 we develop a formulation in terms of auxiliary fermions and derive differential equations for the one- and two-particle irreducible vertex functions under the flow of a sharp frequency cutoff Λ. This method effectively sums up large diagram classes in a systematic way and reaches therefore far beyond the mean-field theory. It turns out that a truncation of the hierarchy of flow equations neglecting all two-loop contributions is not sufficient to adequately describe the competition between order and disorder fluctuations. An improved scheme suggested by Katanin [67] includes certain two-loop terms in the form of self-energy corrections. The latter approach assures that the full (dressed) RPA terms as well as the full particle-particle and particle-hole ladders are contained in the approximation. This proved to be the central requirement to obtain the correct pseudo-fermion damping: While the RPA-diagrams (as the leading term in a $1/S$ expansion) support order tendencies, the ladder diagrams (as the leading term in a $1/N$ expansion, with N being the dimension of the symmetry group SU(N)) favor disorder fluctuations such that altogether the approach allows to treat order and disorder on an equal footing. As a first test, an application to the J_1-J_2 model in fact shows that this approach reproduces the correct phase diagram that has also been obtained by other methods.

Most of the results within that scheme are collected in Chapter 7, where we consider more complicated spin systems. Aside from models based on a square lattice such as the J_1-J_2-J_3 model or the checkerboard model we investigate triangular lattices with real-space anisotropies. Moreover, in order to treat non-Bravais lattices we consider the Heisenberg model on a Kagome- and a honeycomb lattice and finally the Kitaev-Heisenberg model. The latter opens the door to an even larger class of systems as it also involves anisotropies in spin space and ferromagnetic interactions. These systems are adequately described within our FRG method yielding quantum phase-diagrams in good agreement with other numerical studies. While a breakdown of the RG flow of the static $\mathbf{k}$-dependent susceptibility signals magnetic order with wave vector $\mathbf{k}$, the existence of a stable solution indicates the absence of long-range order. This allows us to distinguish between magnetic and non-magnetic phases. Furthermore, unlike other methods, by investigating the $\mathbf{k}$-space resolved susceptibility as the generic outcome of the FRG we can identify the wave vector and the strength of the leading magnetic fluctuations in disordered phases.

To some extent our method allows to characterize the nature of non-magnetic phases: Imposing a small dimer-field perturbation on the system we can calculate the response as a measure for the propensity to form a valence-bond solid. This way we identify parameter regimes of different dimer-fluctuation strength and test different valence-bond patterns against each other, see the corresponding discussions in Sections 6.6 and 7.5. However, for an unambiguous detection of dimer instabilities we need to include the four-particle vertex as the diagrammatic representation of dimer susceptibilities into our RG flow, which turns out to be far beyond the reach of the present formulation.

The natural generalization of our FRG approach to $SU(2)$ broken vertex flows is discussed in Chapter 8. A small external magnetic field breaks the $SU(2)$ symmetry and enables us to calculate the flowing magnetic order parameter. On the basis of an RPA + Hartree approximation the magnetic field regularizes ordering instabilities such that the flow continues into magnetic regimes. Similar findings are obtained in ordered phases using the full one-loop (plus Katanin) scheme. On the other hand, studying the response to external magnetic fields, it turns out to be difficult to detect non-magnetic phases within approximation schemes which do not fulfill conserving properties exactly.

A further generalization of the FRG scheme concerns the implementation of finite temperatures, see Chapter 9. As long as the temperature is not too small we obtain well converged results which give at least a rough indication of the ground-state phase diagram. At finite temperatures it is necessary to employ the Popov-Fedotov scheme to effect exact projection. Comparing the solutions within the exact and the average projection, we obtain the important result that with decreasing temperature the two schemes indeed approach each other which supports our argument that the use of the average projection is justified at zero temperature.

Chapter 10 contains a discussion on zero- and one-dimensional spin systems. It is illustrated that ordering tendencies are overestimated in dimensions lower than two which we regard as a consequence of missing two-loop long-range correlation diagrams. Thus, we conclude that our approach is most powerful in two dimensions.

11 Conclusion and Outlook

To summarize, the work reported here shows that in spite of the fact that quantum spin models are in the strong coupling regime by definition, partial resummations of perturbation theory appear to capture the physics of frustrated magnets. We provide a large body of evidence that our FRG approach is capable (i) of distinguishing between magnetically ordered and disordered phases, (ii) of determining the magnetic wave vector in ordered regimes and the fluctuation profile in disordered regimes and (iii) of estimating and comparing the propensity for the formation of different types of valence-bond solids. In particular, we do not make any assumption on the ground state or perform an expansion around any presumed state. Our starting point of free fermions without dispersion is completely featureless. The resummations, done here in the framework of the FRG method, account in a controlled and systematic way for all two-particle interaction processes, including all couplings between the different channels.

In order to reveal further properties of the systems investigated here it would be most desirable to calculate the magnetic excitation spectrum. However, since our present FRG formulation is based on imaginary Matsubara frequencies, this requires an analytic continuation to the real frequency axis which is known to be an ill-posed problem. Our data on that issue is not yet conclusive enough to be presented in this thesis. Hence, in order to circumvent this problem in future studies, it might be helpful to consider an FRG scheme on the real frequency axis. Another interesting perspective of the spin FRG would be a formulation away from half filling including a finite hopping amplitude of the fermions, i.e., an implementation of the t-J model. We suggest that from both, the methodological as well as from the numerical side such a scheme represents a great challenge. On the other hand, it could provide important insights into the physics of the t-J model, in particular in the context of superconductivity.

11 Conclusion and Outlook

A The Popov-Fedotov Technique

At first glance it is not obvious how the Popov-Fedotov method with its imaginary valued chemical potential projects out the unphysical occupations. In this appendix we want to clarify this issue.

We define Q_i as the operator of the total fermion number on site i. First we note that within the Popov-Fedotov method (i.e, using the Hamiltonian H^{ppv}, see Eq. (3.5)) each particle number Q_i is a conserved quantity. Furthermore, since all particle-number operators mutually commute, an eigenstate of H^{ppv} can be characterized by a set of quantum numbers $\{n, q_1, q_2, \ldots\}$. Here $q_i = 0, 1, 2$ denotes the eigenvalues of the particle-number operator Q_i while n is some set of "physical" quantum numbers in the sectors with all q_i being fixed. The corresponding eigenenergies $E(n, q_1, q_2, \ldots)$ are defined by

$$H|n, q_1, q_2, \ldots\rangle = E(n, q_1, q_2, \ldots)|n, q_1, q_2, \ldots\rangle. \tag{A.1}$$

Note again that H is a Heisenberg Hamiltonian with the spin representation (3.1) already inserted. The physical subspace of H is characterized by the charge configuration with $q_1 = 1$ for all lattice sites i. In this subspace we define the physical eigenenergies by

$$E(n, q_1 = 1, q_2 = 1, \ldots) = E_n. \tag{A.2}$$

A site i with an unphysical occupation carries spin zero. Accordingly a spin operator $\mathbf{S}_i$ annihilates such states,

$$\mathbf{S}_i|n, q_1, q_2, \ldots\rangle = 0 \quad \text{for} \quad q_i = 0 \text{ or } q_i = 2. \tag{A.3}$$

As a consequence, in a subspace with an unphysical auxiliary-particle number at some site i ($q_i = 0$ or $q_i = 2$) the spin at this site is effectively missing in the Hamiltonian H. Therefore, the eigenenergies for $q_i = 0$ and $q_i = 2$ are the same (all other q_i are arbitrary but fixed),

$$E(n, q_1, q_2, \ldots)|_{q_i=0} = E(n, q_1, q_2, \ldots)|_{q_i=2}. \tag{A.4}$$

Now we consider a physical operator $\mathcal{O}$. By "physical" we mean that $\mathcal{O}$ is an arbitrary sum or product of spin operators involving a certain set $\{j\}$ of lattice sites. In the following we show that the expectation value $\langle\mathcal{O}\rangle^{\text{ppv}}$, calculated with H^{ppv} in the entire Hilbert space is identical to the physical expectation value $\langle\mathcal{O}\rangle$ where the average is performed with the original Hamiltonian H and the physical subspace. The expectation value $\langle\mathcal{O}\rangle^{\text{ppv}}$ is given by

$$\langle\mathcal{O}\rangle^{\text{ppv}} = \frac{\text{Tr}^f[e^{-\beta H^{\text{ppv}}}\mathcal{O}]}{\text{Tr}^f[e^{-\beta H^{\text{ppv}}}]}. \tag{A.5}$$

A The Popov-Fedotov Technique

The index f specifies that the trace runs over the enlarged (full) Hilbert space of the fermions. First, we evaluate the partition function Z^{ppv}, i.e., the denominator of Eq. (A.5),

$$
\begin{aligned}
Z^{\mathrm{ppv}} &= \mathrm{Tr}^f[e^{-\beta H^{\mathrm{ppv}}}] \\
&= \sum_n \sum_{q_1,q_2,\ldots} \langle n,q_1,q_2,\ldots | e^{-\beta H^{\mathrm{ppv}}} | n,q_1,q_2,\ldots \rangle \\
&= \sum_n \sum_{q_1,q_2,\ldots} e^{-\beta E(n,q_1,q_2,\ldots)} e^{-i\frac{\pi}{2}q_1} e^{-i\frac{\pi}{2}q_2} \cdots .
\end{aligned} \qquad (\mathrm{A.6})
$$

Here we used the form of H^{ppv} from Eq. (3.5) with $\mu^{\mathrm{ppv}} = -\frac{i\pi}{2\beta}$ and the definition of the eigenenergies, Eq. (A.1). Now we consider the sum over $q_i = 0, 1, 2$ on some site i, while all other particle numbers are fixed: From Eq. (A.4) we know that the contribution of the factor $e^{-\beta E(n,q_1,q_2,\ldots)}$ is the same for $q_i = 0$ and $q_i = 2$, such that the sum $\sum_{q_i=0,2}$ contributes a factor

$$
\sum_{q_i=0,2} e^{-i\frac{\pi}{2}q_i} = \left(1 + e^{-i\pi}\right) = 0 \qquad (\mathrm{A.7})
$$

to the trace, i.e., the unphysical contributions cancel due to the imaginary chemical potential μ^{ppv}. This works for all sites independently, such that only the physical occupation configuration remains,

$$
Z^{\mathrm{ppv}} = \sum_n e^{-\beta E(n,q_1=1,q_2=1,\ldots)} \left(e^{-i\frac{\pi}{2}}\right)^N = \sum_n e^{-\beta E_n} (-i)^N = (-i)^N Z . \qquad (\mathrm{A.8})
$$

Here N is the total number of lattice sites. That is, up to a constant prefactor, the partition function in the enlarged Hilbert space is the physical one.

Now we evaluate the numerator of Eq. (A.5). In addition to the terms in the last line of Eq. (A.6) also the expectation value of the operator $\mathcal{O}$ appears,

$$
\mathrm{Tr}^f[e^{-\beta H^{\mathrm{ppv}}} \mathcal{O}] = \sum_n \sum_{q_1,q_2,\ldots} e^{-\beta E(n,q_1,q_2,\ldots)} e^{-i\frac{\pi}{2}q_1} e^{-i\frac{\pi}{2}q_2} \langle n,q_1,q_2,\ldots | \mathcal{O} | n,q_1,q_2,\ldots \rangle .
$$
(A.9)

For a sum over q_i on a site i which does not belong to the set $\{j\}$ of sites included in $\mathcal{O}$, the argument of Eq. (A.7) holds: The Boltzmann factor as well as the expectation value $\langle n,q_1,q_2,\ldots | \mathcal{O} | n,q_1,q_2,\ldots \rangle$ are the same for $q_i = 0$ and $q_i = 2$, restricting the sum to $q_i = 1$. For a sum over q_i on a site i which belongs to the set $\{j\}$ the restriction to $q_i = 1$ is obvious because due to Eq. (A.3) the expectation value vanishes for $q_i = 0$ and $q_i = 2$. In all cases only physical particle-number contributions remain,

$$
\mathrm{Tr}^f[e^{-\beta H^{\mathrm{ppv}}} \mathcal{O}] = (-i)^N \sum_n e^{-\beta E_n} \langle n,q_1=1,q_2=1,\ldots | \mathcal{O} | n,q_1=1,q_2=1,\ldots \rangle . \qquad (\mathrm{A.10})
$$

The factor $(-i)^N$ appearing in the numerator (A.10) and denominator (A.8) cancels such that the identity $\langle \mathcal{O} \rangle^{\mathrm{ppv}} = \langle \mathcal{O} \rangle$ is proven.

Finally we emphasize that the projection also works for propagators $\langle \mathcal{O}(\tau)\mathcal{O}(0) \rangle$ where the operator acts at different times τ, as long as $\mathcal{O}$ is physical in the sense mentioned above.

B Flow Equations for the Two-Particle Vertex

In this appendix we present the FRG flow-equations for Γ_s^Λ and Γ_d^Λ. After inserting Eq. (6.42a) into Eq. (6.45) and performing the spin sums over α_3 and α_4, we compare the contributions corresponding to spin interactions (proportional to $\sigma^\mu_{\alpha_{1'}\alpha_1}\sigma^\mu_{\alpha_{2'}\alpha_2}$) and density interactions (proportional to $\delta_{\alpha_{1'}\alpha_1}\delta_{\alpha_{2'}\alpha_2}$) on both sides. Furthermore, we use the frequency parametrization introduced in Eqs. (6.38) and (6.39). This leads to the following equations,

$$\frac{d}{d\Lambda}\Gamma^\Lambda_{s\,i_1 i_2}(s,t,u) = \frac{1}{2\pi}\int_{-\infty}^{\infty}d\omega'\Big\{$$
$$\big[-2\Gamma^\Lambda_{s\,i_1 i_2}(s,-\omega_{2'}-\omega',\omega_{1'}+\omega')\Gamma^\Lambda_{s\,i_1 i_2}(s,\omega_2+\omega',\omega_1+\omega')$$
$$+\Gamma^\Lambda_{s\,i_1 i_2}(s,-\omega_{2'}-\omega',\omega_{1'}+\omega')\Gamma^\Lambda_{d\,i_1 i_2}(s,\omega_2+\omega',\omega_1+\omega')$$
$$+\Gamma^\Lambda_{d\,i_1 i_2}(s,-\omega_{2'}-\omega',\omega_{1'}+\omega')\Gamma^\Lambda_{s\,i_1 i_2}(s,\omega_2+\omega',\omega_1+\omega')$$
$$-2\Gamma^\Lambda_{s\,i_1 i_2}(s,\omega_{1'}+\omega',-\omega_{2'}-\omega')\Gamma^\Lambda_{s\,i_1 i_2}(s,-\omega_1-\omega',-\omega_2-\omega')$$
$$+\Gamma^\Lambda_{s\,i_1 i_2}(s,\omega_{1'}+\omega',-\omega_{2'}-\omega')\Gamma^\Lambda_{d\,i_1 i_2}(s,-\omega_1-\omega',-\omega_2-\omega')$$
$$+\Gamma^\Lambda_{d\,i_1 i_2}(s,\omega_{1'}+\omega',-\omega_{2'}-\omega')\Gamma^\Lambda_{s\,i_1 i_2}(s,-\omega_1-\omega',-\omega_2-\omega')\big]$$
$$\times P^\Lambda(\omega',s+\omega')$$
$$+\Big[\,2\sum_j\Gamma^\Lambda_{s\,i_1 j}(\omega_{1'}+\omega',t,\omega_1-\omega')\Gamma^\Lambda_{s\,j i_2}(\omega_2+\omega',t,-\omega_{2'}+\omega')$$
$$+2\sum_j\Gamma^\Lambda_{s\,i_1 j}(\omega_1-\omega',t,\omega_{1'}+\omega')\Gamma^\Lambda_{s\,j i_2}(\omega_{2'}-\omega',t,-\omega_2-\omega')$$
$$+\Gamma^\Lambda_{s\,i_1 i_2}(\omega_{1'}+\omega',t,\omega_1-\omega')\Gamma^\Lambda_{s\,i_2 i_2}(\omega_2+\omega',-\omega_{2'}+\omega',t)$$
$$-\Gamma^\Lambda_{s\,i_1 i_2}(\omega_{1'}+\omega',t,\omega_1-\omega')\Gamma^\Lambda_{d\,i_2 i_2}(\omega_2+\omega',-\omega_{2'}+\omega',t)$$
$$+\Gamma^\Lambda_{s\,i_1 i_2}(\omega_1-\omega',t,\omega_{1'}+\omega')\Gamma^\Lambda_{s\,i_2 i_2}(\omega_{2'}-\omega',-\omega_2-\omega',t)$$
$$-\Gamma^\Lambda_{s\,i_1 i_2}(\omega_1-\omega',t,\omega_{1'}+\omega')\Gamma^\Lambda_{d\,i_2 i_2}(\omega_{2'}-\omega',-\omega_2-\omega',t)$$
$$+\Gamma^\Lambda_{s\,i_1 i_1}(\omega_{1'}+\omega',\omega_1-\omega',t)\Gamma^\Lambda_{s\,i_1 i_2}(\omega_2+\omega',t,-\omega_{2'}+\omega')$$
$$-\Gamma^\Lambda_{d\,i_1 i_1}(\omega_{1'}+\omega',\omega_1-\omega',t)\Gamma^\Lambda_{s\,i_1 i_2}(\omega_2+\omega',t,-\omega_{2'}+\omega')$$
$$+\Gamma^\Lambda_{s\,i_1 i_1}(\omega_1-\omega',\omega_{1'}+\omega',t)\Gamma^\Lambda_{s\,i_1 i_2}(\omega_{2'}-\omega',t,-\omega_2-\omega')$$
$$-\Gamma^\Lambda_{d\,i_1 i_1}(\omega_1-\omega',\omega_{1'}+\omega',t)\Gamma^\Lambda_{s\,i_1 i_2}(\omega_{2'}-\omega',t,-\omega_2-\omega')\Big]$$
$$\times P^\Lambda(\omega',t+\omega')$$

B Flow Equations for the Two-Particle Vertex

$$
\begin{aligned}
-\Big[&\; 2\Gamma^\Lambda_{s\,i_1 i_2}(\omega_{2'}-\omega',-\omega_1-\omega',u)\Gamma^\Lambda_{s\,i_1 i_2}(\omega_2-\omega',\omega_{1'}+\omega',u) \\
&+\Gamma^\Lambda_{s\,i_1 i_2}(\omega_{2'}-\omega',-\omega_1-\omega',u)\Gamma^\Lambda_{d\,i_1 i_2}(\omega_2-\omega',\omega_{1'}+\omega',u) \\
&+\Gamma^\Lambda_{d\,i_1 i_2}(\omega_{2'}-\omega',-\omega_1-\omega',u)\Gamma^\Lambda_{s\,i_1 i_2}(\omega_2-\omega',\omega_{1'}+\omega',u) \\
&+2\Gamma^\Lambda_{s\,i_1 i_2}(\omega_1+\omega',-\omega_{2'}+\omega',u)\Gamma^\Lambda_{s\,i_1 i_2}(\omega_{1'}+\omega',\omega_2-\omega',u) \\
&+\Gamma^\Lambda_{s\,i_1 i_2}(\omega_1+\omega',-\omega_{2'}+\omega',u)\Gamma^\Lambda_{d\,i_1 i_2}(\omega_{1'}+\omega',\omega_2-\omega',u) \\
&+\Gamma^\Lambda_{d\,i_1 i_2}(\omega_1+\omega',-\omega_{2'}+\omega',u)\Gamma^\Lambda_{s\,i_1 i_2}(\omega_{1'}+\omega',\omega_2-\omega',u)\Big] \\
&\times P^\Lambda(\omega',u+\omega')\Big\},
\end{aligned} \quad (B.1)
$$

$$
\begin{aligned}
\frac{d}{d\Lambda}\Gamma^\Lambda_{d\,i_1 i_2}(s,t,u) = \frac{1}{2\pi}\int_{-\infty}^{\infty} d\omega'\Big\{ & \\
\Big[&\; 3\Gamma^\Lambda_{s\,i_1 i_2}(s,-\omega_{2'}-\omega',\omega_{1'}+\omega')\Gamma^\Lambda_{s\,i_1 i_2}(s,\omega_2+\omega',\omega_1+\omega') \\
&+\Gamma^\Lambda_{d\,i_1 i_2}(s,-\omega_{2'}-\omega',\omega_{1'}+\omega')\Gamma^\Lambda_{d\,i_1 i_2}(s,\omega_2+\omega',\omega_1+\omega') \\
&+3\Gamma^\Lambda_{s\,i_1 i_2}(s,\omega_{1'}+\omega',-\omega_{2'}-\omega')\Gamma^\Lambda_{s\,i_1 i_2}(s,-\omega_1-\omega',-\omega_2-\omega') \\
&+\Gamma^\Lambda_{d\,i_1 i_2}(s,\omega_{1'}+\omega',-\omega_{2'}-\omega')\Gamma^\Lambda_{d\,i_1 i_2}(s,-\omega_1-\omega',-\omega_2-\omega')\Big] \\
&\times P^\Lambda(\omega',s+\omega') \\
+\Big[&\; 2\sum_j \Gamma^\Lambda_{d\,i_1 j}(\omega_{1'}+\omega',t,\omega_1-\omega')\Gamma^\Lambda_{d\,j i_2}(\omega_2+\omega',t,-\omega_{2'}+\omega') \\
&+2\sum_j \Gamma^\Lambda_{d\,i_1 j}(\omega_1-\omega',t,\omega_{1'}+\omega')\Gamma^\Lambda_{d\,j i_2}(\omega_{2'}-\omega',t,-\omega_2-\omega') \\
&-3\Gamma^\Lambda_{d\,i_1 i_2}(\omega_{1'}+\omega',t,\omega_1-\omega')\Gamma^\Lambda_{s\,i_2 i_2}(\omega_2+\omega',-\omega_{2'}+\omega',t) \\
&-\Gamma^\Lambda_{d\,i_1 i_2}(\omega_{1'}+\omega',t,\omega_1-\omega')\Gamma^\Lambda_{d\,i_2 i_2}(\omega_2+\omega',-\omega_{2'}+\omega',t) \\
&-3\Gamma^\Lambda_{d\,i_1 i_2}(\omega_1-\omega',t,\omega_{1'}+\omega')\Gamma^\Lambda_{s\,i_2 i_2}(\omega_{2'}-\omega',-\omega_2-\omega',t) \\
&-\Gamma^\Lambda_{d\,i_1 i_2}(\omega_1-\omega',t,\omega_{1'}+\omega')\Gamma^\Lambda_{d\,i_2 i_2}(\omega_{2'}-\omega',-\omega_2-\omega',t) \\
&-3\Gamma^\Lambda_{s\,i_1 i_1}(\omega_{1'}+\omega',\omega_1-\omega',t)\Gamma^\Lambda_{d\,i_1 i_2}(\omega_2+\omega',t,-\omega_{2'}+\omega') \\
&-\Gamma^\Lambda_{d\,i_1 i_1}(\omega_{1'}+\omega',\omega_1-\omega',t)\Gamma^\Lambda_{d\,i_1 i_2}(\omega_2+\omega',t,-\omega_{2'}+\omega') \\
&-3\Gamma^\Lambda_{s\,i_1 i_1}(\omega_1-\omega',\omega_{1'}+\omega',t)\Gamma^\Lambda_{d\,i_1 i_2}(\omega_{2'}-\omega',t,-\omega_2-\omega') \\
&-\Gamma^\Lambda_{d\,i_1 i_1}(\omega_1-\omega',\omega_{1'}+\omega',t)\Gamma^\Lambda_{d\,i_1 i_2}(\omega_{2'}-\omega',t,-\omega_2-\omega')\Big] \\
&\times P^\Lambda(\omega',t+\omega') \\
-\Big[&\; 3\Gamma^\Lambda_{s\,i_1 i_2}(\omega_{2'}-\omega',-\omega_1-\omega',u)\Gamma^\Lambda_{s\,i_1 i_2}(\omega_2-\omega',\omega_{1'}+\omega',u) \\
&+\Gamma^\Lambda_{d\,i_1 i_2}(\omega_{2'}-\omega',-\omega_1-\omega',u)\Gamma^\Lambda_{d\,i_1 i_2}(\omega_2-\omega',\omega_{1'}+\omega',u) \\
&+3\Gamma^\Lambda_{s\,i_1 i_2}(\omega_1+\omega',-\omega_{2'}+\omega',u)\Gamma^\Lambda_{s\,i_1 i_2}(\omega_{1'}+\omega',\omega_2-\omega',u) \\
&+\Gamma^\Lambda_{d\,i_1 i_2}(\omega_1+\omega',-\omega_{2'}+\omega',u)\Gamma^\Lambda_{d\,i_1 i_2}(\omega_{1'}+\omega',\omega_2-\omega',u)\Big] \\
&\times P^\Lambda(\omega',u+\omega')\Big\}.
\end{aligned} \quad (B.2)
$$

B Flow Equations for the Two-Particle Vertex

The frequencies $\omega_{1'}, \omega_{2'}, \omega_1, \omega_2$ on the right hand sides are related to the transfer frequencies via

$$\omega_{1'} = \tfrac{1}{2}(s + t + u), \quad \omega_{2'} = \tfrac{1}{2}(s - t - u),$$
$$\omega_1 = \tfrac{1}{2}(s - t + u), \quad \omega_2 = \tfrac{1}{2}(s + t - u). \quad \text{(B.3)}$$

The definition of $P^\Lambda(\omega_1, \omega_2)$ which denotes a bubble of G^Λ and S^Λ depends on the truncation scheme of the FRG equations. Within the conventional truncation of Section 6.4 the three-particle vertex is completely neglected such that the Green's function G^Λ and the single-scale propagator S^Λ are given by the standard definitions (6.29) and (6.32), respectively. In this scheme $P^\Lambda(\omega_1, \omega_2)$ has the form

$$P^\Lambda(\omega_1, \omega_2) \to P^\Lambda_{\text{con}}(\omega_1, \omega_2) = \frac{\delta(|\omega_1| - \Lambda)\,\Theta(|\omega_2| - \Lambda)}{\omega_1 + \gamma^\Lambda(\omega_1)\,\omega_2 + \gamma^\Lambda(\omega_2)}, \quad \text{(B.4)}$$

and the internal integration $\int d\omega' \ldots$ simplifies to $\sum_{\omega' = \pm \Lambda} \ldots$. Within the Katanin truncation-scheme considered in Section 6.5 the single-scale propagator (as it is used in the second flow equation) acquires an extra term, see Eq. (6.62). In this case we get a more complicated expression,

$$P^\Lambda(\omega_1, \omega_2) \to P^\Lambda_{\text{Kat}}(\omega_1, \omega_2) = \frac{\delta(|\omega_1| - \Lambda)\,\Theta(|\omega_2| - \Lambda)}{\omega_1 + \gamma^\Lambda(\omega_1)\,\omega_2 + \gamma^\Lambda(\omega_2)}$$
$$+ \left(\frac{d}{d\Lambda}\gamma^\Lambda(\omega_1)\right)\frac{\Theta(|\omega_1| - \Lambda)}{(\omega_1 + \gamma^\Lambda(\omega_1))^2}\frac{\Theta(|\omega_2| - \Lambda)}{\omega_2 + \gamma^\Lambda(\omega_2)}. \quad \text{(B.5)}$$

In both schemes $P^\Lambda(\omega_1, \omega_2)$ is an odd function in ω_1 and ω_2 separately. The equations (B.1) and (B.2) may still be simplified as the following consideration shows: Eq. (6.45) can be rewritten such that the terms "$(3 \leftrightarrow 4)$" do not appear, if simultaneously the last line is replaced by "$\times(G^\Lambda(\omega_3)S^\Lambda(\omega_4) + G^\Lambda(\omega_4)S^\Lambda(\omega_3))$". Using this modified form of Eq. (6.45) we end up with the equations

$$\frac{d}{d\Lambda}\Gamma^\Lambda_{s\,i_1 i_2}(s, t, u) = \frac{1}{2\pi}\int_{-\infty}^{\infty} d\omega' \Big\{$$
$$\big[-2\Gamma^\Lambda_{s\,i_1 i_2}(s, -\omega_{2'} - \omega', \omega_{1'} + \omega')\Gamma^\Lambda_{s\,i_1 i_2}(s, \omega_2 + \omega', \omega_1 + \omega')$$
$$+\Gamma^\Lambda_{s\,i_1 i_2}(s, -\omega_{2'} - \omega', \omega_{1'} + \omega')\Gamma^\Lambda_{d\,i_1 i_2}(s, \omega_2 + \omega', \omega_1 + \omega')$$
$$+\Gamma^\Lambda_{d\,i_1 i_2}(s, -\omega_{2'} - \omega', \omega_{1'} + \omega')\Gamma^\Lambda_{s\,i_1 i_2}(s, \omega_2 + \omega', \omega_1 + \omega')\big]$$
$$\times \big[P^\Lambda(\omega', s + \omega') + P^\Lambda(s + \omega', \omega')\big]$$

$$+\Big[2\sum_j \Gamma^\Lambda_{\text{s}\,i_1 j}(\omega_{1'}+\omega',t,\omega_1-\omega')\Gamma^\Lambda_{\text{s}\,j i_2}(\omega_2+\omega',t,-\omega_{2'}+\omega')$$
$$+\Gamma^\Lambda_{\text{s}\,i_1 i_2}(\omega_{1'}+\omega',t,\omega_1-\omega')\Gamma^\Lambda_{\text{s}\,i_1 i_1}(\omega_2+\omega',-\omega_{2'}+\omega',t)$$
$$-\Gamma^\Lambda_{\text{s}\,i_1 i_2}(\omega_{1'}+\omega',t,\omega_1-\omega')\Gamma^\Lambda_{\text{d}\,i_1 i_1}(\omega_2+\omega',-\omega_{2'}+\omega',t)$$
$$+\Gamma^\Lambda_{\text{s}\,i_1 i_1}(\omega_{1'}+\omega',\omega_1-\omega',t)\Gamma^\Lambda_{\text{s}\,i_1 i_2}(\omega_2+\omega',t,-\omega_{2'}+\omega')$$
$$-\Gamma^\Lambda_{\text{d}\,i_1 i_1}(\omega_{1'}+\omega',\omega_1-\omega',t)\Gamma^\Lambda_{\text{s}\,i_1 i_2}(\omega_2+\omega',t,-\omega_{2'}+\omega')\Big]$$
$$\times\big[P^\Lambda(\omega',t+\omega')+P^\Lambda(t+\omega',\omega')\big]$$
$$-\Big[2\Gamma^\Lambda_{\text{s}\,i_1 i_2}(\omega_{2'}-\omega',-\omega_1-\omega',u)\Gamma^\Lambda_{\text{s}\,i_1 i_2}(\omega_2-\omega',\omega_{1'}+\omega',u)$$
$$+\Gamma^\Lambda_{\text{s}\,i_1 i_2}(\omega_{2'}-\omega',-\omega_1-\omega',u)\Gamma^\Lambda_{\text{d}\,i_1 i_2}(\omega_2-\omega',\omega_{1'}+\omega',u)$$
$$+\Gamma^\Lambda_{\text{d}\,i_1 i_2}(\omega_{2'}-\omega',-\omega_1-\omega',u)\Gamma^\Lambda_{\text{s}\,i_1 i_2}(\omega_2-\omega',\omega_{1'}+\omega',u)\Big]$$
$$\times\big[P^\Lambda(\omega',u+\omega')+P^\Lambda(u+\omega',\omega')\big]\Big\}\,,\qquad\text{(B.6)}$$

$$\frac{d}{d\Lambda}\Gamma^\Lambda_{\text{d}\,i_1 i_2}(s,t,u)=\frac{1}{2\pi}\int_{-\infty}^{\infty}d\omega'\Big\{$$
$$\big[3\Gamma^\Lambda_{\text{s}\,i_1 i_2}(s,-\omega_{2'}-\omega',\omega_{1'}+\omega')\Gamma^\Lambda_{\text{s}\,i_1 i_2}(s,\omega_2+\omega',\omega_1+\omega')$$
$$+\Gamma^\Lambda_{\text{d}\,i_1 i_2}(s,-\omega_{2'}-\omega',\omega_{1'}+\omega')\Gamma^\Lambda_{\text{d}\,i_1 i_2}(s,\omega_2+\omega',\omega_1+\omega')\big]$$
$$\times\big[P^\Lambda(\omega',s+\omega')+P^\Lambda(s+\omega',\omega')\big]$$
$$+\Big[2\sum_j \Gamma^\Lambda_{\text{d}\,i_1 j}(\omega_{1'}+\omega',t,\omega_1-\omega')\Gamma^\Lambda_{\text{d}\,j i_2}(\omega_2+\omega',t,-\omega_{2'}+\omega')$$
$$-3\Gamma^\Lambda_{\text{d}\,i_1 i_2}(\omega_{1'}+\omega',t,\omega_1-\omega')\Gamma^\Lambda_{\text{s}\,i_1 i_1}(\omega_2+\omega',-\omega_{2'}+\omega',t)$$
$$-\Gamma^\Lambda_{\text{d}\,i_1 i_2}(\omega_{1'}+\omega',t,\omega_1-\omega')\Gamma^\Lambda_{\text{d}\,i_1 i_1}(\omega_2+\omega',-\omega_{2'}+\omega',t)$$
$$-3\Gamma^\Lambda_{\text{s}\,i_1 i_1}(\omega_{1'}+\omega',\omega_1-\omega',t)\Gamma^\Lambda_{\text{d}\,i_1 i_2}(\omega_2+\omega',t,-\omega_{2'}+\omega')$$
$$-\Gamma^\Lambda_{\text{d}\,i_1 i_1}(\omega_{1'}+\omega',\omega_1-\omega',t)\Gamma^\Lambda_{\text{d}\,i_1 i_2}(\omega_2+\omega',t,-\omega_{2'}+\omega')\Big]$$
$$\times\big[P^\Lambda(\omega',t+\omega')+P^\Lambda(t+\omega',\omega')\big]$$
$$-\Big[3\Gamma^\Lambda_{\text{s}\,i_1 i_2}(\omega_{2'}-\omega',-\omega_1-\omega',u)\Gamma^\Lambda_{\text{s}\,i_1 i_2}(\omega_2-\omega',\omega_{1'}+\omega',u)$$
$$+\Gamma^\Lambda_{\text{d}\,i_1 i_2}(\omega_{2'}-\omega',-\omega_1-\omega',u)\Gamma^\Lambda_{\text{d}\,i_1 i_2}(\omega_2-\omega',\omega_{1'}+\omega',u)\Big]$$
$$\times\big[P^\Lambda(\omega',u+\omega')+P^\Lambda(u+\omega',\omega')\big]\Big\}\,.\qquad\text{(B.7)}$$

Finally, we state the initial conditions which are obtained by comparing Eqs. (6.34) and (6.36),
$$\gamma^{\Lambda\to\infty}(\omega)=0\,,$$
$$\Gamma^{\Lambda\to\infty}_{\text{s}\,i_1 i_2}(s,t,u)=\tfrac{1}{4}J_{i_1 i_2}\,,\qquad \Gamma^{\Lambda\to\infty}_{\text{d}\,i_1 i_2}(s,t,u)=0\,.\qquad\text{(B.8)}$$

C Symmetries of the Two-Particle Vertex in the Transfer Frequencies

We now prove the following symmetries of the two-particle vertex in the SU(2) invariant FRG formulation:
$\Gamma^\Lambda_{s\,i_1 i_2}(s,t,u)$ and $\Gamma^\Lambda_{d\,i_1 i_2}(s,t,u)$ are both invariant under each of the transformations

(1) $s \to -s$, $i_1 \leftrightarrow i_2$,

(2) $t \to -t$,

(3) $u \to -u$, $i_1 \leftrightarrow i_2$,

$\Gamma^\Lambda_{s\,i_1 i_2}(s,t,u)$ is invariant and $\Gamma^\Lambda_{d\,i_1 i_2}(s,t,u)$ changes sign under

(4) $s \leftrightarrow u$.

The transformations (1) and (3) contain an exchange of the sites, $i_1 \leftrightarrow i_2$. For lattices with a monoatomic unit cell (as for example in the case of the J_1-J_2 model), the symmetry under $i_1 \leftrightarrow i_2$ is trivially fulfilled and vertices are invariant under $s \to -s$ and $u \to -u$, individually. For polyatomic lattices (as for example the Kagome lattice) i_1 and i_2 might correspond to different sites in their respective unit cells such that the invariance under $i_1 \leftrightarrow i_2$ is not fulfilled. However, as shown in the following, the invariance under the combined transformations (1) and (3) is intact.

Our starting point is Eq. (6.26b) in integrated form,

$$\Gamma^\Lambda(1',2';1,2) = \Gamma^{\Lambda \to \infty}(1',2';1,2) - \int_\Lambda^\infty d\Lambda'\, [\text{r.h.s. of Eq. (6.26b)}]. \quad (C.1)$$

Generally, a solution of such an equation can be found in an iterative scheme: An initial "guess" $\Gamma^\Lambda_0(1',2';1,2)$ (typically the initial condition) is inserted on the right side. Evaluating the integral, a first approximation $\Gamma^\Lambda_1(1',2';1,2)$ is obtained which will in turn be coupled back to the right side, and so on. The approximations $\Gamma^\Lambda_n(1',2';1,2)$ converge towards the exact solution, i.e., $\Gamma^\Lambda(1',2';1,2) = \Gamma^\Lambda_{n \to \infty}(1',2';1,2)$. Hence, we can prove the above symmetries by complete induction: Assuming that a particular symmetry holds for all vertices on the right side (corresponding to the nth iteration step) we have to show that it is also fulfilled for the vertex on the left side (corresponding to the $(n+1)$th iteration step), provided that the symmetry is already satisfied in the initial solution, i.e., in the initial conditions. This is obviously the case, because $\Gamma^{\Lambda \to \infty}_{s\,i_1 i_2}(s,t,u)$ is constant in all frequencies and $\Gamma^{\Lambda \to \infty}_{d\,i_1 i_2}(s,t,u)$ vanishes. Since FRG treats the variables

C Symmetries of the Two-Particle Vertex in the Transfer Frequencies

$1^{(\prime)}$ and $2^{(\prime)}$ on an equal footing, the two-particle vertex remains unchanged under $1 \leftrightarrow 2$ and $1' \leftrightarrow 2'$, i.e., $\Gamma_m^\Lambda(1', 2'; 1, 2) = \Gamma_m^\Lambda(2', 1'; 2, 1)$ for all m. This identity is often used in the following.

(1) s → −s, i₁ ↔ i₂

Using the frequencies $\omega_{1'}, \omega_{2'}, \omega_1, \omega_2$ the transformation $s \to -s$ corresponds to $\omega_{1'} \to -\omega_{2'}, \omega_{2'} \to -\omega_{1'}, \omega_1 \to -\omega_2, \omega_2 \to -\omega_1$. We introduce the notation $-1 = \{-\omega_1, i_1, \alpha_1\}$ and show that $\Gamma_{n+1}^\Lambda(1', 2'; 1, 2) = \Gamma_{n+1}^\Lambda(-2', -1'; -2, -1)$, assuming that this identity holds in the nth step. For the initial vertex $\Gamma^{\Lambda \to \infty}(1', 2'; 1, 2)$ appearing on the right side of Eq. (C.1) all symmetries are trivially fulfilled. We will not display this term is the following,

$$\Gamma_{n+1}^\Lambda(-2', -1'; -2, -1) = \int_\Lambda^\infty d\Lambda' \frac{1}{2\pi} \sum_{3,4} [\Gamma_n^{\Lambda'}(-2', -1'; 3, 4)\Gamma_n^{\Lambda'}(3, 4; -2, -1)$$

$$- \Gamma_n^{\Lambda'}(-2', 4; -2, 3)\Gamma_n^{\Lambda'}(3, -1'; 4, -1) - (3 \leftrightarrow 4)$$

$$+ \Gamma_n^{\Lambda'}(-1', 4; -2, 3)\Gamma_n^{\Lambda'}(3, -2'; 4, -1) + (3 \leftrightarrow 4)] G^\Lambda(\omega_3) S^\Lambda(\omega_4)$$

$$\stackrel{(I)}{=} \int_\Lambda^\infty d\Lambda' \frac{1}{2\pi} \sum_{3,4} [\Gamma_n^{\Lambda'}(1', 2'; -4, -3)\Gamma_n^{\Lambda'}(-4, -3; 1, 2)$$

$$- \Gamma_n^{\Lambda'}(-4, 2'; -3, 2)\Gamma_n^{\Lambda'}(1', -3; 1, -4) - (3 \leftrightarrow 4)$$

$$+ \Gamma_n^{\Lambda'}(-4, 1'; -3, 2)\Gamma_n^{\Lambda'}(2', -3; 1, -4) + (3 \leftrightarrow 4)] G^\Lambda(\omega_3) S^\Lambda(\omega_4)$$

$$\stackrel{(II)}{=} \int_\Lambda^\infty d\Lambda' \frac{1}{2\pi} \sum_{3,4} [\Gamma_n^{\Lambda'}(1', 2'; 4, 3)\Gamma_n^{\Lambda'}(4, 3; 1, 2)$$

$$- \Gamma_n^{\Lambda'}(4, 2'; 3, 2)\Gamma_n^{\Lambda'}(1', 3; 1, 4) - (3 \leftrightarrow 4)$$

$$+ \Gamma_n^{\Lambda'}(4, 1'; 3, 2)\Gamma_n^{\Lambda'}(2', 3; 1, 4) + (3 \leftrightarrow 4)] G^\Lambda(\omega_3) S^\Lambda(\omega_4)$$

$$\stackrel{(III)}{=} \int_\Lambda^\infty d\Lambda' \frac{1}{2\pi} \sum_{3,4} [\Gamma_n^{\Lambda'}(2', 1'; 3, 4)\Gamma_n^{\Lambda'}(3, 4; 2, 1)$$

$$- \Gamma_n^{\Lambda'}(2', 4; 2, 3)\Gamma_n^{\Lambda'}(3, 1'; 4, 1) - (3 \leftrightarrow 4)$$

$$+ \Gamma_n^{\Lambda'}(1', 4; 2, 3)\Gamma_n^{\Lambda'}(3, 2'; 4, 1) + (3 \leftrightarrow 4)] G^\Lambda(\omega_3) S^\Lambda(\omega_4)$$

$$\stackrel{(IV)}{=} \Gamma_{n+1}^\Lambda(2', 1'; 2, 1)$$

$$\stackrel{(V)}{=} \Gamma_{n+1}^\Lambda(1', 2'; 1, 2). \quad (C.2)$$

Step (I) uses the invariance of $\Gamma_n^\Lambda(1', 2'; 1, 2)$ under $s \to -s$. In step (II) the frequency integrations are transformed by $\omega_3 \to -\omega_3, \omega_4 \to -\omega_4$ and the properties $G^\Lambda(-\omega) = -G^\Lambda(\omega)$ and $S^\Lambda(-\omega) = -S^\Lambda(\omega)$ are employed. The identity $\Gamma^\Lambda(1', 2'; 1, 2) = \Gamma^\Lambda(2', 1'; 2, 1)$ is used in steps (III) and (V).

Now we have proven that the solution fulfills $\Gamma^\Lambda(1', 2'; 1, 2) = \Gamma^\Lambda(-2', -1'; -2, -1)$.

Using the parametrization (6.36) this identity reads

$$\begin{aligned}
\{ & [\Gamma^\Lambda_{s\,i_1 i_2}(\omega_{1'},\omega_{2'};\omega_1,\omega_2)\sigma^\mu_{\alpha_{1'}\alpha_1}\sigma^\mu_{\alpha_{2'}\alpha_2} + \Gamma^\Lambda_{d\,i_1 i_2}(\omega_{1'},\omega_{2'};\omega_1,\omega_2)\delta_{\alpha_{1'}\alpha_1}\delta_{\alpha_{2'}\alpha_2}] \\
& \times \delta_{i_{1'} i_1}\delta_{i_{2'} i_2} \\
& - [\Gamma^\Lambda_{s\,i_1 i_2}(\omega_{1'},\omega_{2'};\omega_2,\omega_1)\sigma^\mu_{\alpha_{1'}\alpha_2}\sigma^\mu_{\alpha_{2'}\alpha_1} + \Gamma^\Lambda_{d\,i_1 i_2}(\omega_{1'},\omega_{2'};\omega_2,\omega_1)\delta_{\alpha_{1'}\alpha_2}\delta_{\alpha_{2'}\alpha_1}] \\
& \times \delta_{i_{1'} i_2}\delta_{i_{2'} i_1} \} \delta(\omega_1+\omega_2-\omega_{1'}-\omega_{2'}) \\
=& \\
\{ & [\Gamma^\Lambda_{s\,i_2 i_1}(-\omega_{2'},-\omega_{1'};-\omega_2,-\omega_1)\sigma^\mu_{\alpha_{2'}\alpha_2}\sigma^\mu_{\alpha_{1'}\alpha_1} + \Gamma^\Lambda_{d\,i_2 i_1}(-\omega_{2'},-\omega_{1'};-\omega_2,-\omega_1)\delta_{\alpha_{2'}\alpha_2}\delta_{\alpha_{1'}\alpha_1}] \\
& \times \delta_{i_{2'} i_2}\delta_{i_{1'} i_1} \\
& - [\Gamma^\Lambda_{s\,i_2 i_1}(-\omega_{2'},-\omega_{1'};-\omega_1,-\omega_2)\sigma^\mu_{\alpha_{2'}\alpha_1}\sigma^\mu_{\alpha_{1'}\alpha_2} + \Gamma^\Lambda_{d\,i_2 i_1}(-\omega_{2'},-\omega_{1'};-\omega_1,-\omega_2)\delta_{\alpha_{2'}\alpha_1}\delta_{\alpha_{1'}\alpha_2}] \\
& \times \delta_{i_{2'} i_1}\delta_{i_{1'} i_2} \} \delta(\omega_1+\omega_2-\omega_{1'}-\omega_{2'}).
\end{aligned} \qquad (C.3)$$

A comparison of terms leads to $\Gamma^\Lambda_{s/d\,i_1 i_2}(\omega_{1'},\omega_{2'};\omega_1,\omega_2) = \Gamma^\Lambda_{s/d\,i_2 i_1}(-\omega_{2'},-\omega_{1'};-\omega_2,-\omega_1)$. Introducing transfer frequencies, the identity $\Gamma^\Lambda_{s/d\,i_1 i_2}(s,t,u) = \Gamma^\Lambda_{s/d\,i_2 i_1}(-s,t,u)$ is proven.

(2) $t \to -t$

The transformation $t \to -t$ corresponds to $\omega_{1'} \leftrightarrow \omega_1$ and $\omega_{2'} \leftrightarrow \omega_2$. We show that $\Gamma^\Lambda_{n+1}(1',2';1,2) = \Gamma^\Lambda_{n+1}(1,2;1',2')$, assuming that this relation holds in the nth iteration step,

$$\begin{aligned}
\Gamma^\Lambda_{n+1}(1,2;1',2') &= \int_\Lambda^\infty d\Lambda' \frac{1}{2\pi}\sum_{3,4}[\Gamma^{\Lambda'}_n(1,2;3,4)\Gamma^{\Lambda'}_n(3,4;1',2') \\
&\quad -\Gamma^{\Lambda'}_n(1,4;1',3)\Gamma^{\Lambda'}_n(3,2;4,2') - (3 \leftrightarrow 4) \\
&\quad +\Gamma^{\Lambda'}_n(2,4;1',3)\Gamma^{\Lambda'}_n(3,1;4,2') + (3 \leftrightarrow 4)]G^\Lambda(\omega_3)S^\Lambda(\omega_4) \\
&\stackrel{(I)}{=} \int_\Lambda^\infty d\Lambda' \frac{1}{2\pi}\sum_{3,4}[\Gamma^{\Lambda'}_n(1',2';3,4)\Gamma^{\Lambda'}_n(3,4;1,2) \\
&\quad -\Gamma^{\Lambda'}_n(1',3;1,4)\Gamma^{\Lambda'}_n(4,2';3,2) - (3 \leftrightarrow 4) \\
&\quad +\Gamma^{\Lambda'}_n(4,2';3,1)\Gamma^{\Lambda'}_n(1',3;2,4) + (3 \leftrightarrow 4)]G^\Lambda(\omega_3)S^\Lambda(\omega_4) \\
&\stackrel{(II)}{=} \int_\Lambda^\infty d\Lambda' \frac{1}{2\pi}\sum_{3,4}[\Gamma^{\Lambda'}_n(1',2';3,4)\Gamma^{\Lambda'}_n(3,4;1,2) \\
&\quad -\Gamma^{\Lambda'}_n(1',4;1,3)\Gamma^{\Lambda'}_n(3,2';4,2) - (3 \leftrightarrow 4) \\
&\quad +\Gamma^{\Lambda'}_n(2',4;1,3)\Gamma^{\Lambda'}_n(3,1';4,2) + (3 \leftrightarrow 4)]G^\Lambda(\omega_3)S^\Lambda(\omega_4) \\
&= \Gamma^\Lambda_{n+1}(1',2';1,2).
\end{aligned} \qquad (C.4)$$

Step (I) uses $\Gamma^\Lambda_n(1',2';1,2) = \Gamma^\Lambda_n(1,2;1',2')$ and interchanges the order of some factors. In step (II) the identity $\Gamma^\Lambda(1',2';1,2) = \Gamma^\Lambda(2',1';2,1)$ is employed in the crossed particle-hole channel. Using the parametrization (6.36) the relation $\Gamma^\Lambda(1',2';1,2) =$

$\Gamma^\Lambda(1, 2; 1', 2')$ becomes

$$\begin{aligned}
\{ & [\Gamma^\Lambda_{s\,i_1 i_2}(\omega_{1'}, \omega_{2'}; \omega_1, \omega_2)\sigma^\mu_{\alpha_{1'}\alpha_1}\sigma^\mu_{\alpha_{2'}\alpha_2} + \Gamma^\Lambda_{d\,i_1 i_2}(\omega_{1'}, \omega_{2'}; \omega_1, \omega_2)\delta_{\alpha_{1'}\alpha_1}\delta_{\alpha_{2'}\alpha_2}] \\
& \times \delta_{i_{1'} i_1}\delta_{i_{2'} i_2} \\
& - [\Gamma^\Lambda_{s\,i_1 i_2}(\omega_{1'}, \omega_{2'}; \omega_2, \omega_1)\sigma^\mu_{\alpha_{1'}\alpha_2}\sigma^\mu_{\alpha_{2'}\alpha_1} + \Gamma^\Lambda_{d\,i_1 i_2}(\omega_{1'}, \omega_{2'}; \omega_2, \omega_1)\delta_{\alpha_{1'}\alpha_2}\delta_{\alpha_{2'}\alpha_1}] \\
& \times \delta_{i_{1'} i_2}\delta_{i_{2'} i_1} \}\delta(\omega_1 + \omega_2 - \omega_{1'} - \omega_{2'}) \\
= \{ & [\Gamma^\Lambda_{s\,i_1 i_2}(\omega_1, \omega_2; \omega_{1'}, \omega_{2'})\sigma^\mu_{\alpha_1\alpha_{1'}}\sigma^\mu_{\alpha_2\alpha_{2'}} + \Gamma^\Lambda_{d\,i_1 i_2}(\omega_1, \omega_2; \omega_{1'}, \omega_{2'})\delta_{\alpha_1\alpha_{1'}}\delta_{\alpha_2\alpha_{2'}}] \\
& \times \delta_{i_1 i_{1'}}\delta_{i_2 i_{2'}} \\
& - [\Gamma^\Lambda_{s\,i_1 i_2}(\omega_1, \omega_2; \omega_{2'}, \omega_{1'})\sigma^\mu_{\alpha_1\alpha_{2'}}\sigma^\mu_{\alpha_2\alpha_{1'}} + \Gamma^\Lambda_{d\,i_1 i_2}(\omega_1, \omega_2; \omega_{2'}, \omega_{1'})\delta_{\alpha_1\alpha_{2'}}\delta_{\alpha_2\alpha_{1'}}] \\
& \times \delta_{i_1 i_{2'}}\delta_{i_2 i_{1'}} \}\delta(\omega_1 + \omega_2 - \omega_{1'} - \omega_{2'}). \quad (C.5)
\end{aligned}$$

Due to $\sigma^\mu_{\alpha_1\alpha_{1'}}\sigma^\mu_{\alpha_2\alpha_{2'}} = \sigma^\mu_{\alpha_{1'}\alpha_1}\sigma^\mu_{\alpha_{2'}\alpha_2}$ the coefficients can also be compared in the spin sector, which finally yields $\Gamma^\Lambda_{s/d\,i_1 i_2}(\omega_{1'}, \omega_{2'}; \omega_1, \omega_2) = \Gamma^\Lambda_{s\,i_1 i_2}(\omega_1, \omega_2; \omega_{1'}, \omega_{2'})$ or equivalently $\Gamma^\Lambda_{s/d\,i_1 i_2}(s, t, u) = \Gamma^\Lambda_{s/d\,i_1 i_2}(s, -t, u)$.

(3) $u \to -u, i_1 \leftrightarrow i_2$

This symmetry is easily proven because we already know that the two-particle vertex is invariant under the exchange of both particles, i.e., $\Gamma^\Lambda(1', 2'; 1, 2) = \Gamma^\Lambda(2', 1'; 2, 1)$. This corresponds to an invariance of Γ^Λ_s and Γ^Λ_d under the combined transformation $t \to -t$, $u \to -u, i_1 \leftrightarrow i_2$. Regarding $t \to -t$ the invariance has already been proven in (2) such that the remaining symmetry under $u \to -u, i_1 \leftrightarrow i_2$ is obvious.

(4) $s \leftrightarrow u$

To prove this invariance we consider the version (6.45) of the second flow equation. For the definition of $\Gamma^\Lambda_=$ see Eq. (6.42a). First we introduce the following notation,

$$\Gamma^\Lambda_{=\,i_1 i_2}(-1', 2'; 1, 2) = \Big[\, \Gamma^\Lambda_{s\,i_1 i_2}(-\omega_{1'}, \omega_{2'}; \omega_1, \omega_2)\sigma^y_{\alpha_{1'}\beta}\sigma^\mu_{\beta\alpha_1}\sigma^\mu_{\alpha_{2'}\alpha_2} \\
+ \Gamma^\Lambda_{d\,i_1 i_2}(-\omega_{1'}, \omega_{2'}; \omega_1, \omega_2)\sigma^y_{\alpha_{1'}\beta}\delta_{\beta\alpha_1}\delta_{\alpha_{2'}\alpha_2}\Big]\delta(\omega_1 + \omega_2 + \omega_{1'} - \omega_{2'}). \quad (C.6)$$

Here, the numbers are composite indices of frequency and spin. Equivalently, the expression $\Gamma^\Lambda_{=\,i_1 i_2}(1', -2'; 1, 2)$ corresponds to the replacements $\omega_{2'} \to -\omega_{2'}$, $\sigma^\mu_{\alpha_{2'}\alpha_2} \to \sigma^y_{\alpha_{2'}\beta}\sigma^\mu_{\beta\alpha_2}$ and $\delta_{\alpha_{2'}\alpha_2} \to \sigma^y_{\alpha_{2'}\beta}\delta_{\beta\alpha_2}$. When the minus sign applies to the third or fourth argument, we use a different convention,

$$\Gamma^\Lambda_{=\,i_1 i_2}(1', 2'; -1, 2) = \Big[\, \Gamma^\Lambda_{s\,i_1 i_2}(\omega_{1'}, \omega_{2'}; -\omega_1, \omega_2)\sigma^\mu_{\alpha_{1'}\beta}\sigma^y_{\beta\alpha_1}\sigma^\mu_{\alpha_{2'}\alpha_2} \\
+ \Gamma^\Lambda_{d\,i_1 i_2}(\omega_{1'}, \omega_{2'}; -\omega_1, \omega_2)\delta_{\alpha_{1'}\beta}\sigma^y_{\beta\alpha_1}\delta_{\alpha_{2'}\alpha_2}\Big]\delta(-\omega_1 + \omega_2 - \omega_{1'} - \omega_{2'}), \quad (C.7)$$

and analog for $\Gamma^\Lambda_{=\,i_1 i_2}(1', 2'; 1, -2)$. Due to $\sigma^y\sigma^y = 1$ this notation is consistent with $\Gamma^\Lambda_{=\,i_1 i_2}(-(-1'), 2'; 1, 2) = \Gamma^\Lambda_{=\,i_1 i_2}(1', 2'; 1, 2)$. We now prove the identity

$$\Gamma^\Lambda_{=\,i_1 i_2}(1', 2'; 1, 2) = -\Gamma^\Lambda_{=\,i_1 i_2}(1', -2; 1, -2'). \quad (C.8)$$

C Symmetries of the Two-Particle Vertex in the Transfer Frequencies

Again we employ the property that the two-particle vertex in invariant under the exchange of both particles, i.e., $\Gamma^\Lambda_{=i_1i_2}(1',2';1,2) = \Gamma^\Lambda_{=i_2i_1}(2',1';2,1)$. The proof is similar as compared to the above invariances. However, to shorten the notation we omit the indices indicating the iteration step. We treat the different terms of Eq. (6.45) individually and start with the particle-particle channel and the crossed particle-hole channel. For the sum of these channels we use the notation $\Gamma^{\Lambda\,\text{a+e}}_{=i_1i_2}(1',2';1,2)$ (where the index "a+e" refers to the labels in Fig. 6.2). For brevity the combination $\frac{1}{2\pi}\int_\Lambda^\infty d\Lambda' \int_{-\infty}^\infty d\omega_4 \int_{-\infty}^\infty d\omega_3 \sum_{\alpha_3\alpha_4}$ appearing in the integrated version of Eq. (6.45) is written as $\widetilde{\sum}$. We transform the quantity $\Gamma^{\Lambda\,\text{a+e}}_{=i_1i_2}(1',-2;1,-2')$ as follows,

$$\begin{aligned}
\Gamma^{\Lambda\,\text{a+e}}_{=i_1i_2}(1',-2;1,-2') &= \widetilde{\sum}[\Gamma^{\Lambda'}_{=i_1i_2}(1',-2;3,4)\Gamma^{\Lambda'}_{=i_1i_2}(3,4;1,-2') + (3\leftrightarrow 4) && \text{(C.9a)}\\
&\quad + \Gamma^{\Lambda'}_{=i_2i_1}(-2,4;3,1)\Gamma^{\Lambda'}_{=i_2i_1}(3,1';-2',4) + (3\leftrightarrow 4)]G^\Lambda(\omega_3)S^\Lambda(\omega_4)\\
&\stackrel{(I)}{=} \widetilde{\sum}[\Gamma^{\Lambda'}_{=i_1i_2}(1',-2;3,4)\Gamma^{\Lambda'}_{=i_1i_2}(3,4;1,-2') + (3\leftrightarrow 4) && \text{(C.9b)}\\
&\quad + \Gamma^{\Lambda'}_{=i_1i_2}(4,-2;1,3)\Gamma^{\Lambda'}_{=i_1i_2}(1',3;4,-2') + (3\leftrightarrow 4)]G^\Lambda(\omega_3)S^\Lambda(\omega_4)\\
&\stackrel{(II)}{=} \widetilde{\sum}[\Gamma^{\Lambda'}_{=i_1i_2}(1',-4;3,2)\Gamma^{\Lambda'}_{=i_1i_2}(3,2';1,-4) + (3\leftrightarrow 4) && \text{(C.9c)}\\
&\quad + \Gamma^{\Lambda'}_{=i_1i_2}(4,-3;1,2)\Gamma^{\Lambda'}_{=i_1i_2}(1',2;4,-3) + (3\leftrightarrow 4)]G^\Lambda(\omega_3)S^\Lambda(\omega_4)\\
&\stackrel{(III)}{=} -\widetilde{\sum}[\Gamma^{\Lambda'}_{=i_1i_2}(1',4;3,2)\Gamma^{\Lambda'}_{=i_1i_2}(3,2';1,4) + (3\leftrightarrow 4) && \text{(C.9d)}\\
&\quad + \Gamma^{\Lambda'}_{=i_1i_2}(4,3;1,2)\Gamma^{\Lambda'}_{=i_1i_2}(1',2;4,3) + (3\leftrightarrow 4)]G^\Lambda(\omega_3)S^\Lambda(\omega_4)\\
&\stackrel{(IV)}{=} -\widetilde{\sum}[\Gamma^{\Lambda'}_{=i_2i_1}(4,1';2,3)\Gamma^{\Lambda'}_{=i_2i_1}(2',3;4,1) + (3\leftrightarrow 4) && \text{(C.9e)}\\
&\quad + \Gamma^{\Lambda'}_{=i_1i_2}(4,3;1,2)\Gamma^{\Lambda'}_{=i_1i_2}(1',2;4,3) + (3\leftrightarrow 4)]G^\Lambda(\omega_3)S^\Lambda(\omega_4)\\
&= -\Gamma^{\Lambda\,\text{a+e}}_{=i_1i_2}(1',2';1,2)\,,
\end{aligned}$$

demonstrating that $s\leftrightarrow u$ interchanges the roles of the particle-particle and the crossed particle-hole channel. In step (I) we exchange particles in the second line of (C.9a) and in (II) we use the assumption (C.8). Thereafter, step (III) transforms variables according to $-3\to 3$ (in the second line of (C.9c)) and $-4\to 4$ (in the first line of (C.9c)). The minus sign appearing in (C.9d) comes from the propagators being odd under these transformations. Finally, in (IV) particles are exchanged in the first line of (C.9d).

Next, we prove the identity (C.8) for the term in Fig. 6.2c,

$$\begin{aligned}
\Gamma^{\Lambda\,\text{c}}_{=i_1i_2}(1',-2;1,-2')&\\
&= -\widetilde{\sum}[\Gamma^{\Lambda'}_{=i_1i_2}(1',4;1,3)\Gamma^{\Lambda'}_{=i_2i_2}(3,-2;-2',4) + (3\leftrightarrow 4)]G^\Lambda(\omega_3)S^\Lambda(\omega_4)\\
&\stackrel{(I)}{=} -\widetilde{\sum}[\Gamma^{\Lambda'}_{=i_1i_2}(1',-3;1,-4)\Gamma^{\Lambda'}_{=i_2i_2}(3,-4;-2',2) + (3\leftrightarrow 4)]G^\Lambda(\omega_3)S^\Lambda(\omega_4)\\
&\stackrel{(II)}{=} -\widetilde{\sum}[\Gamma^{\Lambda'}_{=i_1i_2}(1',-3;1,-4)\Gamma^{\Lambda'}_{=i_2i_2}(-4,3;2,-2') + (3\leftrightarrow 4)]G^\Lambda(\omega_3)S^\Lambda(\omega_4)\\
&\stackrel{(III)}{=} \widetilde{\sum}[\Gamma^{\Lambda'}_{=i_1i_2}(1',-3;1,-4)\Gamma^{\Lambda'}_{=i_2i_2}(-4,2';2,-3) + (3\leftrightarrow 4)]G^\Lambda(\omega_3)S^\Lambda(\omega_4)\\
&\stackrel{(IV)}{=} \widetilde{\sum}[\Gamma^{\Lambda'}_{=i_1i_2}(1',3;1,4)\Gamma^{\Lambda'}_{=i_2i_2}(4,2';2,3) + (3\leftrightarrow 4)]G^\Lambda(\omega_3)S^\Lambda(\omega_4)\\
&= -\Gamma^{\Lambda\,\text{c}}_{=i_1i_2}(1',2';1,2)\,, && \text{(C.10)}
\end{aligned}$$

where we apply the assumption (C.8) to both factors in (I) and to the right factor in (III). In step (II) particles are exchanged and in (IV) we perform the transformations $-3 \to 3$ and $-4 \to 4$ simultaneously.

Finally, the terms in Fig. 6.2b and 6.2d are treated as follows,

$$\begin{aligned}
\Gamma^{\Lambda\, b+d}_{=\,i_1 i_2}(1',-2;1,-2') &= \tilde{\sum}[\sum_j \Gamma^{\Lambda'}_{=\,i_1 j}(1',4;1,3)\Gamma^{\Lambda'}_{=\,j i_2}(3,-2;4,-2') + (3 \leftrightarrow 4) \\
&\quad - \Gamma^{\Lambda'}_{=\,i_1 i_1}(1',4;3,1)\Gamma^{\Lambda'}_{=\,i_1 i_2}(3,-2;4,-2') - (3 \leftrightarrow 4)]G^\Lambda(\omega_3)S^\Lambda(\omega_4) \\
&= -\tilde{\sum}[\sum_j \Gamma^{\Lambda'}_{=\,i_1 j}(1',4;1,3)\Gamma^{\Lambda'}_{=\,j i_2}(3,2';4,2) + (3 \leftrightarrow 4) \\
&\quad - \Gamma^{\Lambda'}_{=\,i_1 i_1}(1',4;3,1)\Gamma^{\Lambda'}_{=\,i_1 i_2}(3,2';4,2) - (3 \leftrightarrow 4)] \\
&= -\Gamma^{\Lambda\, b+d}_{=\,i_1 i_2}(1',2';1,2)\,.
\end{aligned} \tag{C.11}$$

Here, we only employed the assumption (C.8). Having considered all contributions, Eq. (C.8) is now proved. Using the definition (6.42a) together with the Pauli-matrix relations $\sigma^y \sigma^x \sigma^y = -\sigma^x$, $\sigma^y \sigma^z \sigma^y = -\sigma^z$, $(\sigma^x)^T = \sigma^x$, $(\sigma^y)^T = -\sigma^y$ and $(\sigma^z)^T = \sigma^z$ Eq. (C.8) is transformed to

$$\begin{aligned}
&\left[\Gamma^\Lambda_{s\,i_1 i_2}(\omega_{1'},\omega_{2'};\omega_1,\omega_2)\sigma^\mu_{\alpha_{1'}\alpha_1}\sigma^\mu_{\alpha_{2'}\alpha_2} + \Gamma^\Lambda_{d\,i_1 i_2}(\omega_{1'},\omega_{2'};\omega_1,\omega_2)\delta_{\alpha_{1'}\alpha_1}\delta_{\alpha_{2'}\alpha_2}\right] \\
&\times \delta(\omega_1+\omega_2-\omega_{1'}-\omega_{2'}) \\
&= -\left[-\Gamma^\Lambda_{s\,i_1 i_2}(\omega_{1'},-\omega_2;\omega_1,-\omega_{2'})\sigma^\mu_{\alpha_{1'}\alpha_1}\sigma^\mu_{\alpha_{2'}\alpha_2} + \Gamma^\Lambda_{d\,i_1 i_2}(\omega_{1'},-\omega_2;\omega_1,-\omega_{2'})\delta_{\alpha_{1'}\alpha_1}\delta_{\alpha_{2'}\alpha_2}\right] \\
&\times \delta(\omega_1+\omega_2-\omega_{1'}-\omega_{2'})\,.
\end{aligned} \tag{C.12}$$

Comparing the coefficients we obtain

$$\Gamma^\Lambda_{s\,i_1 i_2}(\omega_{1'},\omega_{2'};\omega_1,\omega_2) = \Gamma^\Lambda_{s\,i_1 i_2}(\omega_{1'},-\omega_2;\omega_1,-\omega_{2'})\,, \tag{C.13a}$$

$$\Gamma^\Lambda_{d\,i_1 i_2}(\omega_{1'},\omega_{2'};\omega_1,\omega_2) = -\Gamma^\Lambda_{d\,i_1 i_2}(\omega_{1'},-\omega_2;\omega_1,-\omega_{2'})\,, \tag{C.13b}$$

or expressed in terms of transfer frequencies,

$$\Gamma^\Lambda_{s\,i_1 i_2}(s,t,u) = \Gamma^\Lambda_{s\,i_1 i_2}(u,t,s)\,, \tag{C.14a}$$

$$\Gamma^\Lambda_{d\,i_1 i_2}(s,t,u) = -\Gamma^\Lambda_{d\,i_1 i_2}(u,t,s)\,. \tag{C.14b}$$

This proves the invariance under the transformation (4).

List of Figures

2.1	The J_1-J_2-Heisenberg model	6
4.1	Diagrammatic representation of the Hartree approximation	14
4.2	Phase diagram in the g-T-plane using the Hartree-approximation	16
4.3	Self-consistent RPA equation in diagrammatic form	17
4.4	Dispersion of the spin excitations	19
5.1	Magnetizations within a Hartree approximation assuming a finite pseudo-fermion lifetime	23
5.2	Phase diagram in the γ-g-plane	23
5.3	Static susceptibility within RPA	24
5.4	Dynamical spin structure factor in the paramagnetic phase	26
5.5	Static correlation function	27
5.6	Static correlation length	27
5.7	Dyson's equation for different renormalized propagators	28
6.1	FRG equations for γ_1^Λ and γ_2^Λ in diagrammatic form	36
6.2	Diagrammatic contributions to the two-particle vertex	41
6.3	Diagrammatic contributions to the self energy	41
6.4	Array of 7×7 sites	43
6.5	Phase diagram for a static FRG approximation	49
6.6	Flowing susceptibility for an FRG scheme in the RPA channel	50
6.7	Néel- and collinear susceptibility within the conventional truncation	51
6.8	Additional term within the Katanin truncation	53
6.9	Diagrammatic proof of RPA within the Katanin scheme	55
6.10	Hartree resummation within the Katanin scheme	56
6.11	Flow of the susceptibility within the Katanin scheme	57
6.12	Flow of the Néel susceptibility showing also the unphysical flow	58
6.13	Static susceptibility for the paramagnetic phase	59
6.14	Frequency dependent damping at the end of the FRG-flow	59
6.15	Patterns for columnar and plaquette dimerization	61
6.16	Flowing dimer correlation for $g = 0.55$	61
7.1	Classical phase diagram of the J_1-J_2-J_3 model	64
7.2	Quantum phase diagram of the J_1-J_2-J_3 model	64
7.3	Wave vectors of the dominant susceptibility	65
7.4	Susceptibility for the J_1-J_2-J_3 model in the complete Brillouin zone	66

List of Figures

7.5   The checkerboard lattice   67
7.6   Flow of the Néel susceptibility in the checkerboard model   69
7.7   Susceptibility for the checkerboard model in the extended Brillouin zone   70
7.8   The triangular lattice in real- and reciprocal space   72
7.9 $\mathbf{k}$-space resolved susceptibility for the ATLHM with $\xi \leq 1$ 73
7.10 Susceptibility flow for $\xi = 0$ and $\xi = 1$ 74
7.11 $\mathbf{k}$-space resolved susceptibility for the ATLHM with $\xi \geq 1$ 75
7.12 Susceptibility at the isotropic triangular point $\xi = 1$ 76
7.13 The Kagome model and its static susceptibility   78
7.14 Classical phase diagram of the J_1-J_2-J_3 honeycomb model   79
7.15 FRG phase diagram of the J_1-J_2-J_3 honeycomb model   80
7.16 Evolution of the Heisenberg-honeycomb model along the $J_3 = 0$ axis . . .   81
7.17 Dimer responses along different lines   82
7.18 J_3-sweep of the susceptibility for $J_2 = 0.5$ and $J_2 = 0.6$ 83
7.19 The Kitaev model on the honeycomb lattice   85
7.20 α-sweep of the static magnetic susceptibility   86
7.21 Frustration parameter for the Kitaev-Heisenberg model   89

8.1   RPA + Hartree approximation for the J_1-J_2 model within FRG   95
8.2   Magnetic FRG scheme including all interaction channels   97

9.1   Néel-susceptibility flow within the finite temperature FRG   101
9.2   Néel susceptibility as a function of the temperature   102

10.1 Susceptibility for the one-dimensional J_1-J_2 Heisenberg model   106
10.2 Auxiliary-particle damping for the two-site Heisenberg molecule   106

Acknowledgement

First of all I like to thank my advisor Prof. Dr. Peter Wölfle for his scientific support and for many enlightening discussions. Thanks for providing me such an interesting research topic for my PhD thesis.

I am grateful to Prof. Dr. Alexander Shnirman for co-refereeing this thesis.

A special thanks goes to Ronny Thomale for the fruitful collaboration. Thanks for the numerous phone calls, e-mails and personal discussions, which have been very inspiring for me.

I would like to thank Holger Schmidt for giving me technical insights into the functional renormalization group, especially during the time when I started my work on this method. Thanks also for the good time we had as "Übungsleiter" for "Theorie C für Lehramt".

I thank Peter Schmitteckert for his numerical help and for a first draft of the computer code for the functional renormalization group.

I gratefully acknowledge the help and support of my PhD colleagues Alexander Branschädel, Stefan Kremer, Clemens Müller, Stephane Ngo Dinh, Burkhard Scharfenberger, Michael Schütt and Christian Seiler. Thanks for the stimulating environment you provided.

Finally, I would like to thank all members of the "Institut für Theorie der Kondensierten Materie" for the friendly atmosphere on the 10th floor of the Physikhochhaus. Especially I like to mention Rose Schrempp who is often called the "mother of the institute". Thanks for all the support and the non-scientific activities at the institute.

Bibliography

[1] A. A. Abrikosov, *Electron scattering on magnetic impurities in metals and anomalous resistivity effects*, Physics (Long Island City, N.Y.) **2**, 5 (1965).

[2] A. A. Abrikosov and A. A. Migdal, *On the theory of the Kondo effect*, J. Low Temp. Phys. **3**, 519 (1970).

[3] I. Affleck and J. B. Marston, *Large-n limit of the Heisenberg-Hubbard model: Implications for high-T_c superconductors*, Phys. Rev. B **37**, 3774 (1988).

[4] I. Affleck, Z. Zou, T. Hsu, and P. W. Anderson, *SU(2) gauge symmetry of the large-U limit of the Hubbard model*, Phys. Rev. B **38**, 745 (1988).

[5] S. Andergassen, T. Enss, and V. Meden, *Kondo physics in transport through a quantum dot with Luttinger-liquid leads*, Phys. Rev. B **73**, 153308 (2006).

[6] S. Andergassen, T. Enss, V. Meden, W. Metzner, U. Schollwöck, and K. Schönhammer, *Functional renormalization group for Luttinger liquids with impurities*, Phys. Rev. B **70**, 075102 (2004).

[7] P. W. Anderson, *An approximate quantum theory of the antiferromagnetic ground state*, Phys. Rev. **86**, 694 (1952).

[8] P. W. Anderson, *Resonating valence bonds: A new kind of insulator?*, Mater. Res. Bull. **8**, 153 (1973).

[9] P. W. Anderson, *The resonating valence bond state in La_2CuO_4 and superconductivity*, Science **235**, 1196 (1987).

[10] P. W. Anderson, P. A. Lee, M. Randeria, T. M. Rice, N. Trivedi, and F. C. Zhang, *The physics behind high-temperature superconducting cuprates: the 'plain vanilla' version of RVB*, J. Phys.: Condens. Matter **16**, R755 (2004).

[11] M. Arlego and W. Brenig, *Plaquette order in the J_1-J_2-J_3 model: Series expansion analysis*, Phys. Rev. B **78**, 224415 (2008).

[12] D. P. Arovas and A. Auerbach, *Functional integral theories of low-dimensional quantum Heisenberg models*, Phys. Rev. B **38**, 316 (1988).

[13] A. Auerbach, *Interacting Electrons and Quantum Magnetism* (Springer, 1994).

[14] A. Auerbach and D. P. Arovas, *Spin dynamics in the square-lattice antiferromagnet*, Phys. Rev. Lett. **61**, 617 (1988).

[15] G. Baskaran, S. Mandal, and R. Shankar, *Exact results for spin dynamics and fractionalization in the Kitaev model*, Phys. Rev. Lett. **98**, 247201 (2007).

[16] G. Baskaran, Z. Zou, and P. W. Anderson, *The resonating valence bond state and high-T_c superconductivity - A mean field theory*, Solid State Commun. **63**, 973 (1987).

[17] J. G. Bednorz and K. A. Müller, *Possible high T_c superconductivity in the Ba-La-Cu-O system*, Z. Phys. B **64**, 189 (1986).

[18] E. Berg, E. Altman, and A. Auerbach, *Singlet excitations in pyrochlore: A study of quantum frustration*, Phys. Rev. Lett. **90**, 147204 (2003).

[19] R. F. Bishop, D. J. J. Farnell, and J. B. Parkinson, *Phase transitions in the spin-half J_1-J_2 model*, Phys. Rev. B **58**, 6394 (1998).

[20] R. F. Bishop, P. H. Y. Li, D. J. J. Farnell, and C. E. Campbell, *Magnetic order in a spin-$\frac{1}{2}$ interpolating square-triangle Heisenberg antiferromagnet*, Phys. Rev. B **79**, 174405 (2009).

[21] W. Brenig and A. Honecker, *Planar pyrochlore: A strong-coupling analysis*, Phys. Rev. B **65**, 140407 (2002).

[22] J. Brinckmann and P. A. Lee, *Renormalized mean-field theory of neutron scattering in cuprate superconductors*, Phys. Rev. B **65**, 014502 (2001).

[23] J. Brinckmann and P. Wölfle, *Auxiliary-fermion approach to critical fluctuations in the two-dimensional quantum antiferromagnetic Heisenberg model*, Phys. Rev. B **70**, 174445 (2004).

[24] J. Brinckmann and P. Wölfle, *Description of magnetic short-range order in the 2D Heisenberg model: Auxiliary fermions with reduced self-consistency*, Physica B **359-361**, 798 (2005).

[25] D. C. Cabra, C. A. Lamas, and H. D. Rosales, *Quantum disordered phase on the frustrated honeycomb lattice*, arXiv: 1003.3226, (2010).

[26] B. Canals, *From the square lattice to the checkerboard lattice: Spin-wave and large-n limit analysis*, Phys. Rev. B **65**, 184408 (2002).

[27] L. Capriotti, F. Becca, A. Parola, and S. Sorella, *Resonating valence bond wave functions for strongly frustrated spin systems*, Phys. Rev. Lett. **87**, 097201 (2001).

[28] L. Capriotti, F. Becca, A. Parola, and S. Sorella, *Suppression of dimer correlations in the two-dimensional J_1-J_2 Heisenberg model: An exact diagonalization study*, Phys. Rev. B **67**, 212402 (2003).

[29] L. Capriotti and S. Sachdev, *Low-temperature broken-symmetry phases of spiral antiferromagnets*, Phys. Rev. Lett. **93**, 257206 (2004).

[30] L. Capriotti and S. Sorella, *Spontaneous plaquette dimerization in the J_1-J_2 Heisenberg model*, Phys. Rev. Lett. **84**, 3173 (2000).

[31] L. Capriotti, A. E. Trumper, and S. Sorella, *Long-range Néel order in the triangular Heisenberg model*, Phys. Rev. Lett. **82**, 3899 (1999).

[32] P. Carretta, R. Melzi, N. Papinutto, and P. Millet, *Very-low-frequency excitations in frustrated two-dimensional $S = 1/2$ Heisenberg antiferromagnets*, Phys. Rev. Lett. **88**, 047601 (2002).

[33] H. A. Ceccatto, C. J. Gazza, and A. E. Trumper, *Nonclassical disordered phase in the strong quantum limit of frustrated antiferromagnets*, Phys. Rev. B **47**, 12329 (1993).

[34] J. Chaloupka, G. Jackeli, and G. Khaliullin, *Kitaev-Heisenberg model on a honeycomb lattice: Possible exotic phases in iridium oxides A_2IrO_3*, Phys. Rev. Lett. **105**, 027204 (2010).

[35] P. Chandra, P. Coleman, and A. I. Larkin, *Ising transition in frustrated Heisenberg models*, Phys. Rev. Lett. **64**, 88 (1990).

[36] P. Chandra and B. Doucot, *Possible spin-liquid state at large S for the frustrated square Heisenberg lattice*, Phys. Rev. B **38**, 9335 (1988).

[37] C. W. Chu, P. H. Hor, R. L. Meng, L. Gao, Z. J. Huang, and Y. Q. Wang, *Evidence for superconductivity above 40 K in the La-Ba-Cu-O compound system*, Phys. Rev. Lett. **58**, 405 (1987).

[38] B. K. Clark, D. A. Abanin, and S. L. Sondhi, *Nature of the spin liquid state of the Hubbard model on honeycomb lattice*, arXiv: 1010.3011, (2010).

[39] R. Coldea, D. A. Tennant, A. M. Tsvelik, and Z. Tylczynski, *Experimental realization of a 2D fractional quantum spin liquid*, Phys. Rev. Lett. **86**, 1335 (2001).

[40] P. Coleman, *New approach to the mixed-valence problem*, Phys. Rev. B **29**, 3035 (1984).

[41] E. Dagotto and A. Moreo, *Phase diagram of the frustrated spin-1/2 Heisenberg antiferromagnet in two dimensions*, Phys. Rev. Lett. **63**, 2148 (1989).

[42] J. Dai, Q. Si, J. X. Zhu, and E. Abrahams, *Iron pnictides as a new setting for quantum criticality*, Proc. Natl. Acad. Sci. U.S.A. **106**, 4118 (2009).

[43] R. Darradi, O. Derzhko, R. Zinke, J. Schulenburg, S. E. Krüger, and J. Richter, *Ground state phases of the spin-1/2 J_1-J_2 Heisenberg antiferromagnet on the square lattice: A high-order coupled cluster treatment*, Phys. Rev. B **78**, 214415 (2008).

[44] R. Dillenschneider and J. Richert, *Site-occupation constraints in mean-field approaches to quantum spin systems at finite temperature*, Phys. Rev. B **73**, 024409 (2006).

[45] A. V. Dotsenko and O. P. Sushkov, *Quantum phase transition in the frustrated Heisenberg antiferromagnet*, Phys. Rev. B **50**, 13821 (1994).

[46] T. Einarsson and H. J. Schulz, *Direct calculation of the spin stiffness in the J_1-J_2 Heisenberg antiferromagnet*, Phys. Rev. B **51**, 6151 (1995).

[47] J. Ferrer, *Spin-liquid phase for the frustrated quantum Heisenberg antiferromagnet on a square lattice*, Phys. Rev. B **47**, 8769 (1993).

[48] J. B. Fouet, M. Mambrini, P. Sindzingre, and C. Lhuillier, *Planar pyrochlore: A valence-bond crystal*, Phys. Rev. B **67**, 054411 (2003).

[49] J. B. Fouet, P. Sindzingre, and C. Lhuillier, *An investigation of the quantum J_1-J_2-J_3 model on the honeycomb lattice*, Eur. Phys. J. B **20**, 241 (2001).

[50] M. P. Gelfand, *Series investigations of magnetically disordered ground states in two-dimensional frustrated quantum antiferromagnets*, Phys. Rev. B **42**, 8206 (1990).

[51] R. Gersch, C. Honerkamp, and W. Metzner, *Superconductivity in the attractive Hubbard model: functional renormalization group analysis*, New J. Phys. **10**, 045003 (2008).

[52] R. Gersch, C. Honerkamp, D. Rohe, and W. Metzner, *Fermionic renormalization group flow into phases with broken discrete symmetry: charge-density wave mean-field model*, Eur. Phys. J. B **48**, 349 (2005).

[53] C. J. Halboth and W. Metzner, *Renormalization-group analysis of the two-dimensional Hubbard model*, Phys. Rev. B **61**, 7364 (2000).

[54] R. Hedden, V. Meden, T. Pruschke, and K. Schönhammer, *A functional renormalization group approach to zero-dimensional interacting systems*, J. Phys.: Condens. Matter **16**, 5279 (2004).

[55] W. Heisenberg, *Zur Theorie des Ferromagnetismus*, Zeits. f. Physik **49**, 619 (1928).

[56] J. S. Helton, K. Matan, M. P. Shores, E. A. Nytko, B. M. Bartlett, Y. Yoshida, Y. Takano, A. Suslov, Y. Qiu, J. H. Chung, D. G. Nocera, and Y. S. Lee, *Spin dynamics of the spin-1/2 Kagome lattice antiferromagnet $ZnCu_3(OH)_6Cl_2$*, Phys. Rev. Lett. **98**, 107204 (2007).

[57] C. L. Henley, *Ordering due to disorder in a frustrated vector antiferromagnet*, Phys. Rev. Lett. **62**, 2056 (1989).

[58] C. Honerkamp, M. Salmhofer, N. Furukawa, and T. M. Rice, *Breakdown of the Landau-Fermi liquid in two dimensions due to umklapp scattering*, Phys. Rev. B **63**, 035109 (2001).

[59] C. Husemann and M. Salmhofer, *Efficient parametrization of the vertex function, Ω scheme, and the t, t' Hubbard model at van Hove filling*, Phys. Rev. B **79**, 195125 (2009).

[60] M. Inui, S. Doniach, and M. Gabay, *Doping dependence of antiferromagnetic correlations in high-temperature superconductors*, Phys. Rev. B **38**, 6631 (1988).

[61] M. Inui, S. Doniach, P. J. Hirschfeld, and A. E. Ruckenstein, *Coexistence of antiferromagnetism and superconductivity in a mean-field theory of high-T_c superconductors*, Phys. Rev. B **37**, 2320 (1988).

[62] L. Isaev, G. Ortiz, and J. Dukelsky, *Hierarchical mean-field approach to the J_1-J_2 Heisenberg model on a square lattice*, Phys. Rev. B **79**, 024409 (2009).

[63] T. Itou, A. Oyamada, S. Maegawa, M. Tamura, and R. Kato., *Quantum spin liquid in the spin-1/2 triangular antiferromagnet $EtMe_3Sb[Pd(dmit)_2]_2$*, Phys. Rev. B **77**, 104413 (2008).

[64] H. C. Jiang, Z. Y. Weng, and D. N. Sheng, *Density matrix renormalization group numerical study of the kagome antiferromagnet*, Phys. Rev. Lett. **101**, 117203 (2008).

[65] C. Karrasch, T. Enss, and V. Meden, *Functional renormalization group approach to transport through correlated quantum dots*, Phys. Rev. B **73**, 235337 (2006).

[66] C. Karrasch, R. Hedden, R. Peters, T. Pruschke, K. Schönhammer, and V. Meden, *A finite-frequency functional renormalization group approach to the single impurity Anderson model*, J. Phys.: Condens. Matter **20**, 345205 (2008).

[67] A. A. Katanin, *Fulfillment of Ward identities in the functional renormalization group approach*, Phys. Rev. B **70**, 115109 (2004).

[68] G. Khaliullin and P. Horsch, *Doping dependence of long-range magnetic order in the t-J model*, Phys. Rev. B **47**, 463 (1993).

[69] M. N. Kiselev and R. Oppermann, *Spin-glass transition in a Kondo lattice with quenched disorder*, JETP Lett. **71**, 250 (2000).

[70] A. Kitaev, *Anyons in an exactly solved model and beyond*, Ann. Phys. **321**, 2 (2006).

[71] A. M. Läuchli and C. Lhuillier, *Dynamical correlations of the kagome $S = 1/2$ Heisenberg quantum antiferromagnet*, arXiv: 0901.1065, (2009).

[72] P. W. Leung and N. W. Lam, *Numerical evidence for the spin-Peierls state in the frustrated quantum antiferromagnet*, Phys. Rev. B **53**, 2213 (1996).

[73] P. W. Leung and K. J. Runge, *Spin-1/2 quantum antiferromagnets on the triangular lattice*, Phys. Rev. B **47**, 5861 (1993).

[74] M. Mambrini, A. Läuchli, D. Poilblanc, and F. Mila, *Plaquette valence-bond crystal in the frustrated Heisenberg quantum antiferromagnet on the square lattice*, Phys. Rev. B **74**, 144422 (2006).

[75] E. Manousakis, *The spin-1/2 Heisenberg antiferromagnet on a square lattice and its application to the cuprous oxides*, Rev. Mod. Phys. **63**, 1 (1991).

[76] J. B. Marston and I. Affleck, *Large-n limit of the Hubbard-Heisenberg model*, Phys. Rev. B **39**, 11538 (1989).

[77] B. T. Matthias, R. M. Bozorth, and J. H. V. Vleck, *Ferromagnetic interaction in EuO*, Phys. Rev. Lett. **7**, 160 (1961).

[78] R. Melzi, P. Carretta, A. Lascialfari, M. Mambrini, M. Troyer, P. Millet, and F. Mila, *$Li_2VO(Si,Ge)O_4$, a prototype of a two-dimensional frustrated quantum Heisenberg antiferromagnet*, Phys. Rev. Lett. **85**, 1318 (2000).

[79] Z. Y. Meng, T. C. Lang, S. Wessel, F. F. Assaad, and A. Muramatsu, *Quantum spin liquid emerging in two-dimensional correlated Dirac fermions*, Nature **464**, 847 (2010).

[80] W. Metzner, *Functional renormalization group computation of interacting Fermi systems*, Prog. Theor. Phys. Supplement **160**, 58 (2005).

[81] F. Mezzacapo and J. I. Cirac, *Ground-state properties of the spin- antiferromagnetic Heisenberg model on the triangular lattice: a variational study based on entangled-plaquette states*, New J. Phys. **12**, 103039 (2010).

[82] T. R. Morris, *The exact renormalization group and approximate solutions*, Int. J. Mod. Phys. A **9**, 2411 (1994).

[83] H. Mosadeq, F. Shahbazi, and S. A. Jafari, *Plaquette valence-bond ordering in J_1-J_2 Heisenberg antiferromagnet on the honeycomb lattice*, arXiv: 1007.0127, (2010).

[84] A. Mulder, R. Ganesh, L. Capriotti, and A. Paramekanti, *Spiral order by disorder and lattice nematic order in a frustrated Heisenberg antiferromagnet on the honeycomb lattice*, Phys. Rev. B **81**, 214419 (2010).

[85] V. Murg, F. Verstraete, and J. I. Cirac, *Exploring frustrated spin systems using projected entangled pair states*, Phys. Rev. B **79**, 195119 (2009).

[86] J. W. Negele and H. Orland, *Quantum Many-Particle Systems* (Addison-Wesley, Reading, MA, 1988).

[87] A. A. Nersesyan and A. M. Tsvelik, *Spinons in more than one dimension: Resonance valence bond state stabilized by frustration*, Phys. Rev. B **67**, 024422 (2003).

[88] K. Okamoto and K. Nomura, *Fluid-dimer critical point in $S = \frac{1}{2}$ antiferromagnetic Heisenberg chain with next nearest neighbor interactions*, Phys. Lett. A **169**, 433 (1992).

[89] S. E. Palmer and J. T. Chalker, *Quantum disorder in the two-dimensional pyrochlore Heisenberg antiferromagnet*, Phys. Rev. B **64**, 094412 (2001).

[90] T. Pardini and R. R. P. Singh, *Magnetic order in coupled spin-half and spin-one Heisenberg chains in an anisotropic triangular-lattice geometry*, Phys. Rev. B **77**, 214433 (2008).

[91] V. N. Popov and S. A. Fedotov, *The functional-integration method and diagram technique for spin systems*, Sov. Phys. JETP **67**, 535 (1988).

[92] A. Ralko, M. Mambrini, and D. Poilblanc, *Generalized quantum dimer model applied to the frustrated Heisenberg model on the square lattice: Emergence of a mixed columnar-plaquette phase*, Phys. Rev. B **80**, 184427 (2009).

[93] N. Read and S. Sachdev, *Large-N expansion for frustrated quantum antiferromagnets*, Phys. Rev. Lett. **66**, 1773 (1991).

[94] J. D. Reger and A. P. Young, *Monte Carlo simulations of the spin-1/2 Heisenberg antiferromagnet on a square lattice*, Phys. Rev. B **37**, 5978 (1988).

[95] J. Reiss, D. Rohe, and W. Metzner, *Renormalized mean-field analysis of antiferromagnetism and d-wave superconductivity in the two-dimensional Hubbard model*, Phys. Rev. B **75**, 075110 (2007).

[96] J. Reuther, D. A. Abanin, and R. Thomale, *Magnetic order and paramagnetic phases in the quantum J_1-J_2-J_3 honeycomb model*, arXiv: 1103.0859, (2011).

[97] J. Reuther and R. Thomale, *Functional renormalization group for the anisotropic triangular antiferromagnet*, Phys. Rev. B **83**, 024402 (2011).

[98] J. Reuther and P. Wölfle, *J_1-J_2 frustrated two-dimensional Heisenberg model: Random phase approximation and functional renormalization group*, Phys. Rev. B **81**, 144410 (2010).

[99] J. Reuther, P. Wölfle, R. Darradi, W. Brenig, M. Arlego, and J. Richter, *Quantum phases of the planar antiferromagnetic J_1-J_2-J_3 Heisenberg model*, Phys. Rev. B **83**, 064416 (2011).

[100] G. Rickayzen, *Green's Functions and Condensed Matter* (Academic Press, New York, 1980).

[101] D. S. Rokhsar and S. A. Kivelson, *Superconductivity and the quantum hard-core dimer gas*, Phys. Rev. Lett. **61**, 2376 (1988).

[102] A. E. Ruckenstein, P. J. Hirschfeld, and J. Appel, *Mean-field theory of high-T_c superconductivity: The superexchange mechanism*, Phys. Rev. B **36**, 857 (1987).

[103] S. Ryu, O. I. Motrunich, J. Alicea, and M. P. A. Fisher, *Algebraic vortex liquid theory of a quantum antiferromagnet on the kagome lattice*, Phys. Rev. B **75**, 184406 (2007).

[104] S. Sachdev, *Quantum Phase Transitions* (Cambridge University Press, 1999).

[105] S. Sachdev and R. N. Bhatt, *Bond-operator representation of quantum spins: Mean-field theory of frustrated quantum Heisenberg antiferromagnets*, Phys. Rev. B **41**, 9323 (1990).

[106] M. Salmhofer and C. Honerkamp, *Fermionic renormalization group flows*, Prog. Theor. Phys. **105**, 1 (2001).

[107] M. Salmhofer, C. Honerkamp, W. Metzner, and O. Lauscher, *Renormalization group flows into phases with broken symmetry*, Prog. Theor. Phys. **112**, 943 (2004).

[108] H. Schmidt and P. Wölfle, *Transport through a Kondo quantum dot: Functional RG approach*, Ann. Phys. (Berlin) **19**, 60 (2009).

[109] H. J. Schulz, T. A. L. Ziman, and D. Poilblanc, *Magnetic order and disorder in the frustrated quantum Heisenberg antiferromagnet in two dimensions*, J. Phys. I **6**, 675 (1996).

[110] T. Senthil, L. Balents, S. Sachdev, A. Vishwanath, and M. P. A. Fisher, *Quantum criticality beyond the Landau-Ginzburg-Wilson paradigm*, Phys. Rev. B **70**, 144407 (2004).

[111] T. Senthil, A. Vishwanath, L. Balents, S. Sachdev, and M. P. A. Fisher, *Deconfined quantum critical points*, Science **303**, 1490 (2004).

[112] S. Q. Shen and F. C. Zhang, *Antiferromagnetic Heisenberg model on an anisotropic triangular lattice in the presence of a magnetic field*, Phys. Rev. B **66**, 172407 (2002).

[113] Y. Shimizu, K. Miyagawa, K. Kanoda, M. Maesato, and G. Saito, *Spin liquid state in an organic Mott insulator with a triangular lattice*, Phys. Rev. Lett. **91**, 107001 (2003).

[114] Q. Si and E. Abrahams, *Strong correlations and magnetic frustration in the high T_c iron pnictides*, Phys. Rev. Lett. **101**, 076401 (2008).

[115] Q. Si, E. Abrahams, J. Dai, and J. X. Zhu, *Correlation effects in the iron pnictides*, New J. Phys. **11**, 045001 (2009).

[116] P. Sindzingre, *Spin-1/2 frustrated antiferromagnet on a spatially anisotropic square lattice: Contribution of exact diagonalizations*, Phys. Rev. B **69**, 094418 (2004).

[117] P. Sindzingre, J. B. Fouet, and C. Lhuillier, *One-dimensional behavior and sliding Luttinger liquid phase in a frustrated spin-$\frac{1}{2}$ crossed chain model: Contribution of exact diagonalizations*, Phys. Rev. B **66**, 174424 (2002).

[118] P. Sindzingre and C. Lhuillier, *Low-energy excitations of the kagomé antiferromagnet and the spin-gap issue*, EPL **88**, 27009 (2009).

[119] P. Sindzingre, N. Shannon, and T. Momoi, *Phase diagram of the spin-1/2 J_1-J_2-J_3 Heisenberg model on the square lattice*, J. Phys.: Conf. Ser. **200**, 022058 (2010).

[120] R. R. P. Singh and D. A. Huse, *Three-sublattice order in triangular- and Kagomé-lattice spin-half antiferromagnets*, Phys. Rev. Lett. **68**, 1766 (1992).

[121] R. R. P. Singh and D. A. Huse, *Ground state of the spin-1/2 kagome-lattice Heisenberg antiferromagnet*, Phys. Rev. B **76**, 180407 (2007).

[122] R. R. P. Singh, O. A. Starykh, and P. J. Freitas, *A new paradigm for two-dimensional spin liquids*, J. Appl. Phys. **83**, 7387 (1998).

[123] R. R. P. Singh, Z. Weihong, C. J. Hamer, and J. Oitmaa, *Dimer order with striped correlations in the J_1-J_2 Heisenberg model*, Phys. Rev. B **60**, 7278 (1999).

[124] Y. Singh and P. Gegenwart, *Antiferromagnetic Mott insulating state in single crystals of the hexagonal lattice material Na_2IrO_3*, arXiv: 1006.0437, (2010).

[125] J. Sirker, Z. Weihong, O. P. Sushkov, and J. Oitmaa, *J_1-J_2 model: First-order phase transition versus deconfinement of spinons*, Phys. Rev. B **73**, 184420 (2006).

[126] L. Siurakshina, D. Ihle, and R. Hayn, *Magnetic order and finite-temperature properties of the two-dimensional frustrated Heisenberg model*, Phys. Rev. B **64**, 104406 (2001).

[127] O. A. Starykh and L. Balents, *Dimerized phase and transitions in a spatially anisotropic square lattice antiferromagnet*, Phys. Rev. Lett. **93**, 127202 (2004).

[128] O. A. Starykh and L. Balents, *Ordering in spatially anisotropic triangular antiferromagnets*, Phys. Rev. Lett. **98**, 077205 (2007).

[129] O. A. Starykh, A. Furusaki, and L. Balents, *Anisotropic pyrochlores and the global phase diagram of the checkerboard antiferromagnet*, Phys. Rev. B **72**, 094416 (2005).

[130] O. A. Starykh, R. R. P. Singh, and G. C. Levine, *Spinons in a crossed-chains model of a 2D spin liquid*, Phys. Rev. Lett. **88**, 167203 (2002).

[131] O. P. Sushkov, J. Oitmaa, and Z. Weihong, *Quantum phase transitions in the two-dimensional J_1-J_2 model*, Phys. Rev. B **63**, 104420 (2001).

[132] O. P. Sushkov, J. Oitmaa, and Z. Weihong, *Critical dynamics of singlet and triplet excitations in strongly frustrated spin systems*, Phys. Rev. B **66**, 054401 (2002).

[133] D. T. Teaney, M. J. Freiser, and R. W. H. Stevenson, *Discovery of a simple cubic antiferromagnet: Antiferromagnetic resonance in $RbMnF_3$*, Phys. Rev. Lett. **9**, 212 (1962).

[134] T. R. Thurston, R. J. Birgeneau, M. A. Kastner, N. W. Preyer, G. Shirane, Y. Fujii, K. Yamada, Y. Endoh, K. Kakurai, M. Matsuda, Y. Hidaka, and T. Murakami, *Neutron scattering study of the magnetic excitations in metallic and superconducting $La_{2-x}Sr_xCuO_{4-y}$*, Phys. Rev. B **40**, 4585 (1989).

[135] M. U. Ubbens and P. A. Lee, *Flux phases in the t-J model*, Phys. Rev. B **46**, 8434 (1992).

[136] C. Waldtmann, H. U. Everts, B. Bernu, C. Lhuillier, P. Sindzingre, P. Lecheminant, and L. Pierre, *First excitations of the spin 1/2 Heisenberg antiferromagnet on the kagomé lattice*, Eur. Phys. J. B **2**, 501 (1998).

[137] L. Wang, K. S. D. Beach, and A. W. Sandvik, *High-precision finite-size scaling analysis of the quantum-critical point of $S = 1\!2$ Heisenberg antiferromagnetic bilayers*, Phys. Rev. B **73**, 014431 (2006).

[138] Y. R. Wang, *Low-dimensional quantum antiferromagnetic Heisenberg model studied using Wigner-Jordan transformations*, Phys. Rev. B **46**, 151 (1992).

[139] Z. Weihong, *Various series expansions for the bilayer $S = \frac{1}{2}$ Heisenberg antiferromagnet*, Phys. Rev. B **55**, 12267 (1997).

[140] Z. Weihong, R. H. McKenzie, and R. R. P. Singh, *Phase diagram for a class of spin-$\frac{1}{2}$ Heisenberg models interpolating between the square-lattice, the triangular-lattice, and the linear-chain limits*, Phys. Rev. B **59**, 14367 (1999).

[141] M. Q. Weng, D. N. Sheng, Z. Y. Weng, and R. J. Bursill, *Spin-liquid phase in an anisotropic triangular-lattice Heisenberg model: Exact diagonalization and density-matrix renormalization group calculations*, Phys. Rev. B **74**, 012407 (2006).

[142] C. Wetterich, *Exact evolution equation for the effective potential*, Phys. Lett. B **301**, 90 (1993).

[143] D.-K. Yu, Q. Gu, H.-T. Wang, and J.-L. Shen, *Bond-operator approach to the bilayer Heisenberg antiferromagnet*, Phys. Rev. B **59**, 111 (1999).

[144] S. Yunoki and S. Sorella, *Two spin liquid phases in the spatially anisotropic triangular Heisenberg model*, Phys. Rev. B **74**, 014408 (2006).

[145] G. M. Zhang, Y. H. Su, Z. Y. Lu, Z. Y. Weng, D. H. Lee, and T. Xiang, *Universal linear-temperature dependence of static magnetic susceptibility in iron pnictides*, EPL **86**, 37006 (2009).

[146] M. E. Zhitomirsky and K. Ueda, *Valence-bond crystal phase of a frustrated spin-1/2 square-lattice antiferromagnet*, Phys. Rev. B **54**, 9007 (1996).

Die VDM Verlagsservicegesellschaft sucht für wissenschaftliche Verlage abgeschlossene und herausragende

Dissertationen, Habilitationen, Diplomarbeiten, Master Theses, Magisterarbeiten usw.

für die kostenlose Publikation als Fachbuch.

Sie verfügen über eine Arbeit, die hohen inhaltlichen und formalen Ansprüchen genügt, und haben Interesse an einer honorarvergüteten Publikation?

Dann senden Sie bitte erste Informationen über sich und Ihre Arbeit per Email an *info@vdm-vsg.de*.

Sie erhalten kurzfristig unser Feedback!

VDM Verlagsservicegesellschaft mbH
Dudweiler Landstr. 99   Telefon +49 681 3720 174
D - 66123 Saarbrücken   Fax   +49 681 3720 1749
www.vdm-vsg.de

Die VDM Verlagsservicegesellschaft mbH vertritt

Printed by Books on Demand GmbH, Norderstedt / Germany